POTATO
Production Processing and Marketing

POTATO
Production Processing and Marketing

by
Sharad Chandra Yadav

BIOTECH
2008

Biotech Books
Delhi - 110 035

ISBN10 81-7622-198-8
ISBN13 978-81-7622-198-6

Published by : **BIOTECH BOOKS**
1123/74, Tri Nagar,
Delhi - 110 035
Phone: 27383999
e-mail: biotechbooks@yahoo.co.in

Showroom : 4762-63/23, Ansari Road, Darya Ganj,
New Delhi - 110 002
Phone: 23245578, 23244987

Printed at : **Chawla Offset Printers**
New Delhi - 110 052

PRINTED IN INDIA

Preface

Potatoes, *Solanum tubersum,* are grown in almost every country and form a significant part of the diet in whole world. They are the fourth-largest world crop, surpassed in total production only by wheat, rice and corn, a rich course of carbohydrates, potatoes, along with a diet of milk, can supply all the nutrients the human body needs. As such, potato is a major food crop and their production is essential to maintain the supply of one of the world's foremost food staples.

The present text looks into the production techniques, crop processing and marketing processes that are involved in the supply of the reliable, uncomplicated crop that is the potato. Apart from delineating the various species and breeds of potato, the text gives a comprehensive account of the food value uses, cultivation techniques, production, processing, revenues, etc., which are wired in the potato produce. In addition, it undertakes a comprehensive look into various breeding techniques which are being increasingly used to improve the quality and quantity of potatoes, as well as technology which is bringing dramatic changes into the way potato production is executed and managed. In addition to covering current trends and developments in the field, the book also devotes space to understanding the future of the business of potato production.

Sharad Chandra Yadav

Contents

1

Introduction

Potato is the world's most widely grown tuber crop, and the fourth largest food crop in terms of fresh produce - after rice, wheat, and maize (corn). The potato originated in the area of contemporary Peru and Bolivia, identified more specifically in research published by David Spooner in 2005 as an area of southern Peru, just north of Lake Titicaca. The potato was introduced to Europe around 1700, and subsequently by European mariners to territories and ports throughout the world. Thousands of varieties persist in the Andes, where over 100 varieties might be found in a single valley, and a dozen or more might be maintained by a single agricultural household. However, today, over 99% of all cultivated potato varieties worldwide are descendants of a subspecies indigenous to Chile.

Once established in Europe, the potato soon became an important food staple and field crop. Lack of genetic diversity, due to the fact that very few varieties were initially introduced, left the crop vulnerable to disease. In 1845, a fungal disease, Phytophthora infestans, also known as late blight, spread rapidly through the poorer communities of western Ireland, resulting in the Great Irish Famine.

The potato is also strongly associated with Idaho, Maine, Washington, North Dakota, Prince Edward Island, Ireland, Jersey and Russia because of its large role in the agricultural economy and history of these regions. But in recent decades, the greatest expansion of potato has been in Asia, where as

of 2007 approximately eighty percent of the world potato crop is grown. Since the dissolution of the Soviet Union, China has become the world's largest potato producer, followed by India. Potato plants grow high to the ground and bear yellow to silver flowers with yellow stamens.

Potatoes are cross-pollinated mostly by bumblebees that carry pollen from other potato plants, but a substantial amount of self-fertilising occurs as well. Any potato variety can also be propagated vegetatively by planting tubers, pieces of tubers, cut to include at least one or two eyes, or also by cuttings, a practice used in greenhouses for the production of healthy seed tubers. Some commercial potato varieties do not produce seeds at all and are propagated only from tuber pieces. Confusingly, these tubers or tuber pieces are called "seed potatoes".

After potato plants flower, some varieties will produce small green fruits that resemble green cherry tomatoes, each containing up to 300 true seeds. By finely chopping the fruit and soaking it in water, the seeds will separate from the flesh by sinking to the bottom after about a day. All new potato varieties are grown from seeds, also called "true seed" or "botanical seed" to distinguish it from seed tubers. Potato fruit contains large amounts of the toxic alkaloid solanine, and is therefore unsuitable for consumption.

There is general agreement among contemporary botanists that the potato originated in the Andes, all the way from Colombia and Venezuela to Chile and northern Argentina, but with a concentration of genetic diversity, both in the form of cultivated and wild species, in the area of modern day Peru. The evidence thus far shows that the potato was first cultivated in Peru some 7,000 years ago.

The potatoes cultivated in the Andes are not all the same species. However, the major species grown worldwide is Solanum tuberosum. Modern varieties of this species are the most widely cultivated worldwide. There are also four diploid species: Solanum stenotomum, Solanum phureja, Solanum goniocalyx and Solanum ajanhuiri. There are two triploid

species: Solanum chaucha and Solanum juzepczukii. There is one pentaploid cultivated species: Solanum curtilobum.

There are two major subspecies of Solanum tuberosum: andigena, or Andean; and tuberosum, or Chilean. The Andean potato is adapted to the short-day conditions prevalent in the mountainous equatorial and tropical regions where it originated. The Chilean potato is adapted to the long-day conditions prevalent in the higher latitude region of southern Chile, especially on Chiloe Island where it is thought to have originated.

The Andean Solanum tuberosum andigena was the variety first introduced to Europe by way of the Canary Islands and Spain as early as 1700, and dominated European production until a few decades before the Irish Potato Famine, according to recently-published DNA analysis. The same research shows that in the early 19th century, the Chilean Solanum tuberosum tuberosum, adapted to long-day growing conditions, was introduced to Europe. It quickly replaced the Andean short-day variety. Today, over 99% of all cultivated potato varieties worldwide are descendants of the Chilean subspecies.

Historical and genetic evidence suggests that the potato reached India not very much later than Europe, taken there by either the British or the Portuguese. Genetic studies show that all 32 varieties of potato grown in India derive from the Chilean subspecies. The earliest unequivocal reference to the potato in India is in an 1847 British journal. There are about five thousand potato varieties world wide. Three thousand of them are found in the Andes alone, mainly in Peru, Bolivia, Ecuador, Chile and Colombia. They belong to eight or nine species, depending on the taxonomic school.

Apart from the five thousand cultivated varieties, there are about 200 wild species and subspecies, many of which can be crossbred with cultivated varieties, which has been done repeatedly to transfer resistances to certain pests and diseases from the gene pool of wild species to the gene pool of cultivated potato species.

Potatoes are occasionally referred to as "Irish" potatoes in the English speaking world because in the earliest years, this signified a distinction from sweet potatoes. Potatoes were widely grown in Ireland after they were introduced, but in 1845, potato blight devastated the crop, precipitating the six-year-long Great Irish Famine. Most modern potatoes grown in North America arrived through European settlement and not independently from the South American sources. However, at least one wild potato species, Solanum fendleri, is found as far north as Texas and used in breeding for resistance to a nematode species attacking cultivated potatoes.

A secondary center of genetic variability of the potato is Mexico, where important wild species are found that have been used extensively in modern breeding, such as the hexaploid Solanum demissum, as a source of resistance to the devastating late blight disease.

The potato became an important staple crop in northern Europe as the climate changed due to the Little Ice Age, when traditional crops in this region did not produce as reliably as before. At times when and where most other crops would fail, potatoes could still typically be relied upon to contribute adequately to food supplies during the colder years.

The potato was not popular in France during this time, and it is believed that some of the infamous famines could have been lessened if French farmers had adopted it. Today, the potato forms an important part of the traditional cuisines of most of Europe. Belarus has the highest consumption of potato per capita with each Belarusian consuming 338 kg in 2005.

Nutrition

Nutritionally, potatoes are best known for their carbohydrate content. Starch is the predominant form of carbohydrate found in potatoes. A small but significant portion of the starch in potatoes is resistant to enzymatic digestion in the stomach and small intestine and, thus, reaches the large intestine essentially intact. This resistant starch is considered to have similar physiological effects and health benefits of fiber.

The amount of resistant starch found in potatoes is highly dependent upon preparation methods. Cooking and then cooling potatoes significantly increases resistant starch. For example, cooked potato starch contains about 7% resistant starch, which increases to about 13% upon cooling.

Potato, raw, with peel Nutritional value per 100 g (3.5 oz)	
Energy 80 kcal 320 kJ	
Carbohydrates	19 g
- Starch 15 g	
- Dietary fiber 2.2 g	
Fat	0.1 g
Protein	2 g
Water	75 g
Thiamin (Vit. B1) 0.08 mg	6%
Riboflavin (Vit. B2) 0.03 mg	2%
Niacin (Vit. B3) 1.1 mg	7%
Vitamin B6 0.25 mg	19%
Vitamin C 20 mg	33%
Calcium 12 mg	1%
Iron 1.8 mg	14%
Magnesium 23 mg	6%
Phosphorus 57 mg	8%
Potassium 421 mg	9%
Sodium 6 mg	0%

Potatoes contain a number of important vitamins and minerals. A medium potato (150g/5.3 oz) with the skin provides 27 mg vitamin C (45% of the Daily Value (DV)), 620 mg of potassium (18% of DV), 0.2 mg vitamin B6 (10% of DV) and trace amounts of thiamin, riboflavin, folate, niacin, magnesium, phosphorus, iron, and zinc. Moreover, the fiber content of a potato with skin (2 grams) equals that of many whole grain breads, pastas, and cereals. In addition to vitamins, minerals and fiber, potatoes also contain an assortment of phytochemicals, such as carotenoids and polyphenols. The notion that "all of the potato's nutrients" are

found in the skin is an urban legend. While the skin does contain approximately half of the total dietary fiber, the majority (more than 50%) of the nutrients are found within the potato itself. The cooking method used can significantly impact the nutrient availability of the potato.

Potatoes are often broadly classified as "high" on the Glycemic index (GI) and thus are frequently excluded from the diets of individuals trying to follow a "low GI" eating regimen. In fact, the GI of potatoes can vary considerably depending on the type, origin, preparation methods, and with what it is consumed. Potatoes are prepared in many ways: skin-on or peeled, whole or cut up, with seasonings or without. The only requirement involves cooking to break down the starch. Most potato dishes are served hot, but some are first cooked then served cold, notably potato salad and potato chips/crisps.

Common dishes are: mashed potatoes, which are first boiled (usually peeled), and then mashed with milk or yogurt and butter; whole baked potatoes; boiled or steamed potatoes; French-fried potatoes or chips; cut into cubes and roasted; scalloped, diced, or sliced and fried (home fries); grated into small thin strips and fried (hash browns); grated and formed into dumplings, Rösti or potato pancakes. Unlike many foods, potatoes can also be easily cooked in a microwave oven and still retain nearly all of their nutritional value, provided that they are covered in ventilated plastic wrap to prevent moisture from escaping-this method produces a meal very similar to a steamed potato while retaining the appearance of a conventionally baked potato. Potato chunks also commonly appear as a stew ingredient. Potatoes are boiled between 10 and 25 minutes, depending on size and type, to become soft.

Regional Dishes

Latin America

Peruvian Cuisine naturally contains the potato as a primary ingredient in many dishes, as around 3,000 varieties of this tuber are grown there. Some of the more famous dishes

include Papa a la huancaina, Papa rellena, Ocopa, Carapulcra, Causa and Cau Cau among many others. In Ecuador the potato, as well as being a staple with most dishes, is featured in the hearty Locro de Papas, a thick soup of potato, squash, and cheese.

Europe

In Britain potatoes form part of the traditional staple fish and chips. Mashed, potatoes also form a major component of several other traditional British dishes such as shepherd's pie, bubble and squeak, champ and the 'mashit tatties' (Scots language) which accompany haggis. The Tattie scone is another popular Scottish dish containing potatoes. They are also often sautéed to accompany a meal. Colcannon is a traditional Irish dish involving mashed potato combined with shredded cabbage and onion. Boxty pancakes are eaten all over Ireland, although associated especially with the north, and in Irish diaspora communities: they are traditionally made with grated potatoes, soaked to loosen the starch and mixed with flour, buttermilk and baking powder. A variant eaten and sold in Lancashire, especially Liverpool, is made with cooked and mashed potatoes.

In Northern Europe, especially Denmark, Norway, Sweden and Finland, newly harvested, early ripening varieties are considered a special delicacy. Boiled whole and served with dill, these "new potatoes" are traditionally consumed together with Baltic herring. In the UK, new potatoes are typically cooked with mint and served with a little melted butter - Jersey Royal potatoes are the most prized new potatoes, and have their own Protected Designation of Origin.

In Western Europe, especially in Belgium, sliced potatoes are fried to get frieten, the original French fried potatoes. Potatoes are very popular in continental Europe as well. In Italy, they serve to make a type of pasta called gnocchi. In western Norway, komle is popular. A traditional Canary Islands dish is Canarian wrinkly potatoes or Papas arrugadas.

Tortilla de patatas and Patatas bravas are near-universal constituent of Spanish tapas.

North America

In the United States, potatoes have become one of the most widely consumed crops, and thus have a variety of preparation methods and condiments. French fries and often hash browns are commonly found in typical American fast-food burger joints and cafeterias. One popular favorite involves a baked potato with cheddar cheese on top, and in New England "smashed potatoes" have great popularity. Potato flakes are popular as an instant variety of mashed potatoes, which reconstitute into mashed potatoes by adding water, plus butter & salt for taste. A regional dish of Central New York, salt potatoes are bite-sized new potatoes boiled in water saturated with salt then served with melted butter.

A traditional Acadian dish from New Brunswick is known as poutine râpée. The Acadian poutine is a ball of grated and mashed potato, salted, sometimes filled with pork in the center, and boiled. The result is a moist ball about the size of a baseball. It is commonly eaten with salt and pepper or brown sugar. It is believed to have originated from the German Kloße, prepared by early German settlers who lived among the Acadians. Poutine, by contrast, is a hearty serving of french fries, fresh cheese curds and hot gravy. Tracing its origins to Quebec in the 1950s, it has become popular across Canada and can usually be found where Canadians gather abroad.

Potatoes Toxic Compounds

Potatoes contain glycoalkaloids, toxic compounds, of which the most prevalent are solanine and chaconine. Cooking at high temperatures (over 170°C or 340°F) partly destroys these. The concentration of glycoalkaloid in wild potatoes suffices to produce toxic effects in humans. Glycoalkaloids occur in the greatest concentrations just underneath the skin of the tuber, and they increase with age and exposure to light. Glycoalk-aloids may cause headaches, diarrhea, cramps and in severe

cases coma and death; however, poisoning from potatoes occurs very rarely.

Light exposure also causes greening, thus giving a visual clue as to areas of the tuber that may have become more toxic; however, this does not provide a definitive guide, as greening and glycoalkaloid accumulation can occur independently of each other. Some varieties of potato contain greater glycoalkaloid concentrations than others; breeders developing new varieties test for this, and sometimes have to discard an otherwise promising cultivar.

Breeders try to keep solanine levels below 200 mg/kg (200 ppmw). However, when these commercial varieties turn green, even they can approach concentrations of solanine of 1000 mg/kg (1000 ppmw). In normal potatoes, however, analysis has shown solanine levels may be as little as 3.5% of the breeders' maximum, with 7-187 mg/kg being found. The National Toxicology Programme suggests that the average American consumes at most 12.5 mg/day of solanine from potatoes. Dr. Douglas L. Holt, the State Extension Specialist for Food Safety at the University of Missouri - Columbia, notes that no reported cases of potato-source solanine poisoning have occurred in the U.S. in the last 50 years and most cases involved eating green potatoes or drinking potato-leaf tea.

Solanine is also found in other plants, mainly in the mostly-deadly nightshade family, which includes a minority of edible plants including the potato and the tomato, and other typically more dangerous plants like tobacco. This poison affects the nervous system causing weakness and confusion.

Cultivation Process

Potatoes are generally grown from the eyes of another potato and not from seed. Home gardeners often plant a piece of potato with two or three eyes in a hill of mounded soil. Commercial growers plant potatoes as a row crop using seed tubers, young plants or microtubers and may mound the entire row. At harvest time, gardeners generally dig up potatoes with a long-handled, three-prong "grape" (or graip),

i.e. a spading fork, or a potato hook which is similar to the graip, except the tines are at a 90 degree angle to the handle as is the blade of a hoe. In larger plots, the plow can serve as the most expeditious implement for unearthing potatoes. Commercial harvesting is typically done with large potato harvesters which scoop up the plant and the surrounding earth. This is transported up an apron chain consisting of steel links several feet wide, which separates some of the dirt.

The chain deposits into an area where further separation occurs. Different designs employ different systems at this point. The most complex designs use vine choppers and shakers, along with a blower system or "Flying Willard" to separate the potatoes from the plant. The result is then usually run past workers who continue to sort out plant material, stones, and rotten potatoes before the potatoes are continuously delivered to a wagon or truck. Further inspection and separation occurs when the potatoes are unloaded from the field vehicles and put into storage.

Correct potato husbandry is an arduous task in the best of circumstances. Good ground preparation, harrowing, plowing, and rolling are always needed, along with a little grace from the weather and a good source of water. Three successive plowings, with associated harrowing and rolling, are desirable before planting. Eliminating all root-weeds is desirable in potato cultivation. Potatoes are the most fruitful of the root crops, but much care and consideration is needed to keep them satisfied and fruitful. It is important to harvest potatoes before heavy frosts begin, since field frost damages potatoes in the ground, and even cold weather makes potatoes more susceptible to bruising and possibly later rotting which can quickly ruin a large stored crop.

Seed potato crops are 'rogued' in some countries to eliminate diseased plants or those of a different variety from the seed crop.Storage facilities need to be carefully designed to keep the potatoes alive and slow the natural process of decomposition, which involves the breakdown of starch. It is crucial that the storage area is dark, well ventilated and for long-term storage maintained at temperatures near 40°F (4°C).

For short-term storage prior to cooking, temperatures of about 45-50°F (7-10°C) are preferred. Temperatures below 40°F (4°C) convert potatoes' starch into sugar, which alters their taste and cooking qualities and leads to higher acrylamide levels in the cooked product, especially in deep-fried dishes.

Under optimum conditions possible in commercial warehouses, potatoes can be stored for up to six months, but several weeks is the normal shelf life in homes. If potatoes develop green areas or start to sprout, these areas should be trimmed before using. FAO reports that the world production of potatoes in 2005 was 319 million tonnes. The largest producer, China, accounted for one-fourth of the global output followed by Russia and India.

Role in World Food Production

In 2008, several international organisations began to give more emphasis to the potato as a key part of world food production, due to several developing economic problems. They cited the potato's potential for a beneficial role in world food production, owing to its status as a cheap and plentiful crop which can be raised in a wide variety of climates and locales.

Varieties

Potatoes have been bred into many standard or well-known varieties, each of which have particular agricultural or culinary attributes. Varieties are generally categorised into a few main groups, such as Russets, Reds, Whites, Yellows (aka Yukons), and Purples based on common characteristics. Popular varieties found in markets may include:

- Désirée
- Bintje
- Fianna
- King Edward
- Kipfler
- New
- Nicola

- — Pink Eye
- — Pink Fir Apple
- — Red Pontiac
- — Rooster
- — Russet Burbank
- — Spunta
- — Chiloe·
  - — Camota
  - — Cielo
  - — Pachacoña
  - — Clavela Blanca
  - — Cabritas
  - — Chelina

Genetic research on the potato has resulted in at least one genetically-modified variety, the New Leaf, owned by Monsanto corporation. Potatoes of all varieties are generally cured after harvest to thicken the skin. Prior to curing, the skin is very thin and delicate. These potatoes are sometimes sold as "New Potatoes" and are partic-ularly flavorful. New potatoes are often harvested by the home gardener or farmer by "grabbling", i.e. pulling out the young tubers by hand while leaving the plant in place. In additions, markets may sometimes present various thin-skinned potato varieties as "new potatoes". Some horticulturists sell chimeras, made by grafting a tomato plant onto a potato plant, producing both edible tomatoes and potatoes. This practice is not very widespread.

References

Brook, R. C., R. J. Fick, and T. D. Forbush, *Potato Storage Design and Managemen*, East Lansing, MI.: Michigan State University Agricultural Engineering Information Series, AEIS#627, 1995.

Christ, B. J., *Potato Diseases in Pennsylvania,* Pennsylvania College of Agricultural Sciences, University Park, PA. 1998.

2

Growing of Potato Plants

The potato tuber is an enlarged portion of an underground stem or stollen. Tuber eyes are the buds from which next season's growth will emerge. Eyes are concentrated near the apical end of the tuber, with fewer near the stollen or basal end. Eye number and distribution are characteristic of the variety. The tuber skin is composed of two layers of cells: an outer layer of single cells called the epidermis, underlain by several layers of corky cells called the periderm.

The cells in the periderm layer may contain a pigment that produces coloured potatoes. Below the periderm is the cortex, followed by the vascular ring, which contains the cells that transport food products to the tuber from the above ground stems. Internal to the vascular ring is the medulla. The medulla represents the primary storage area for the potato tuber. Excess food produced by the potato plant is transported to the medulla via the vascular tissue. Cells in the medulla increase in number and size as they are supplied with food, causing the tuber to increase in size.

Growth and quality of potatoes are influenced by environmental factors such as temperature, moisture, light, soil type and nutrients. Many factors that influence potato growth are largely uncontrollable: length of growing season, air and soil temperatures, light intensity and duration, humidity and wind.

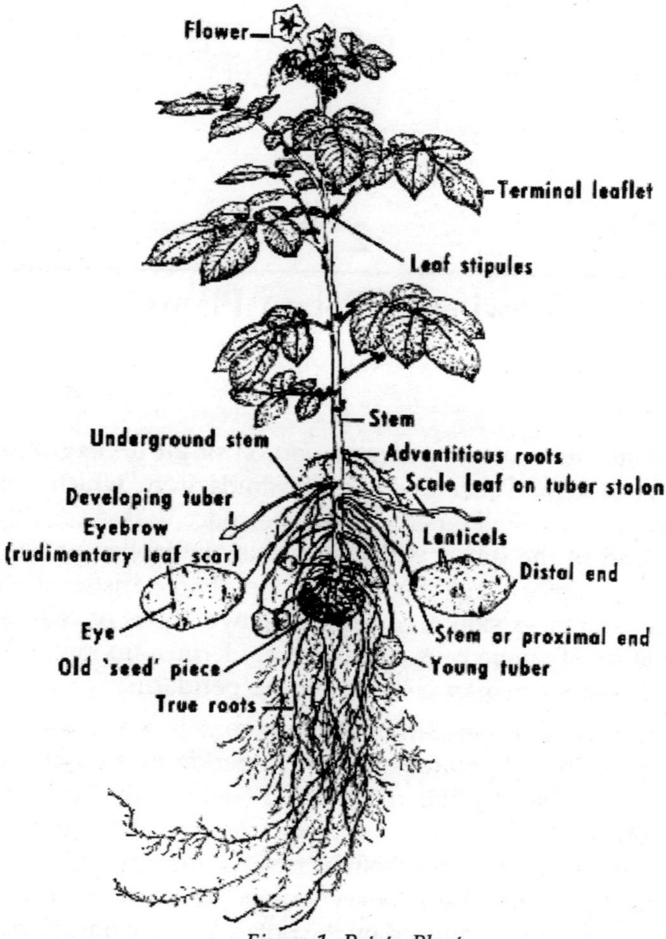

Figure 1. Potato Plant

Other factors that influence growth of the crop can be controlled by the grower: variety of potato, size of mother seed tubers, seed-piece cutting, seed-piece types, cut-seed size, planter operation, plant stand, stem population, moisture, nutrition, pest management, planting date and harvest date. Only when all factors are at optimum levels can the most profitable yields of quality potatoes be attained.

LENGTH OF GROWING SEASON

The potato is a cool-season crop that grows well in certain areas of the Prairie Provinces. The rate of development of sprouts from seed pieces depends on soil temperature. Very little sprout elongation occurs at 43° F (6°C), elongation is slow at 48° F (9°C) and is maximised at about 64°F (18°C). The optimum soil temperature for initiating tubers is 61-66° F (16-19°C). Tuber development declines as soil temperatures rise above 68° F (20°C) and tuber growth practically stops at soil temperatures above 86°F (30°C). The number of tubers set per plant is greater at lower temperatures than at higher temperatures, whereas higher temperatures favour development of large tubers.

Yields are highest when average daytime temperatures are about 69°F (21°C). Cool night tempe-ratures are important because they affect the accumulation of carbohydrates and dry matter in the tubers. At lower night temperatures, respiration is slowed, which enhances storage of starch in the tubers. Physiological Days (P-Days) are a measure of the heat useful for the growth and development of potatoes. In the potato production areas of the Prairies, the highest average accumulation of P-Days, 850-950, occurs south of Lake Manitoba to the U.S. border and the lowest accumulation, 750-900, occurs in the seed production areas south of Edmonton and in Central Saskatchewan.

The varieties currently grown on the Prairies require anywhere from 800-1000 P-Days to reach full maturity. An early-maturing variety such as Norland requires 800 P-Days, and later maturing varieties such as Russet Burbank require 1000 P-Days. Growing a potato variety in an area with insufficient P-Days will reduce yield and affect tuber quality factors such as accumulation of solids and fry colour. The average accumulation of P-Days is insufficient in many parts of the Prairies to produce late season varieties where the crop must be fully mature before harvest. These areas may be suitable for seed or table production, where the crop is killed or harvested before full maturity.

Moisture Levels

Potatoes require a continuous supply of soil water along with adequate soil aeration. Yields are greatest when soil moisture is maintained above 65% of the available soil water (ASW) capacity Tuber set is particularly sensitive to moisture stress. There are generally fewer tubers set when available soil moisture is maintained below 65% of the available soil water (ASW) capacity. The amount of water needed by potatoes varies with the soil type, temperature, humidity, air movement, plant and stem populations, variety and cultural practices. In Alberta and Saskatchewan, too little moisture and fluctuating moisture levels are more common than excessive moisture.

In Manitoba, the situation is reversed. Too much moisture generally causes more problems than too little moisture. Low or fluctuating moisture levels can contribute to common scab, early dying, hollow heart, knobby tubers, low dry matter, low tuber set, and low yield. Excessive soil moisture resulting in poor aeration and water logging of the soil reduces yields and in extreme cases causes tuber rot. An excess of moisture may also lead to enlarged lenticels, which are openings of the epidermis. This detracts from their appearance and allows entry of disease organisms, causing tuber rot in storage.

Growth Stages

The development of potatoes can be broken down into five distinct growth stages.

Growth Stage I: Sprout Development

This stage begins with sprouts developing from the eyes and ends at emergence from the soil. The seed piece is the sole energy source for growth during this stage.

Growth Stage II: Vegetative Growth

This stage, in which all vegetative parts of the plants (leaves, branches, roots and stolons) are formed, begins at emergence and lasts until tubers start to develop. Growth stages I and II

last from 30 to 70 days depending on planting date, soil temperature and other environmental factors, the physiological age of the tubers, and the characteristics of particular cultivators.

Growth Stage III: Tuber Set/Initiation

Tubers are forming at salon tips, but are not yet enlarging. This stage will last approximately 2 weeks.

Growth Stage IV: Tuber Bulking

Tuber cells expand with the accumulation of water, nutrients and carbohydrates. Tuber bulking is the growth stage of longest duration. Depending on date of planting and cultivar, bulking can last up to three months.

Growth Stage V: Maturation

Vines turn yellow and lose leaves, photosynthesis gradually decreases, tuber growth rate slows and the vines die. This stage may not occur when growing a long season variety like Russet Burbank in a production area with a short growing season.

Growth Stage III tuber initiation or tuberisation starts when the ends of stolons begin to swell. This occurs between early and late June depending on location, planting date, climate, soil type, and variety. Tubers form when the plant produces more carbohydrates than are required for vine growth. Varying weather and moisture conditions cause uneven tuber set and growth. The number of tubers formed per plant is called the tuber set. The plant may initially produce 20 to 30 small tubers, but only 5 to 15 tubers typically reach maturity. The growing plant absorbs some of the tubers in the original set.

The number of tubers that achieve maturity is related to available moisture and nutrition. Optimum moisture and nutrient levels early in the growing season are critical to the maintenance and development of tubers. Maintaining available soil moisture above 65% of the available soil water (ASW) capacity has been shown to produce high tuber set

when compared to lower moisture levels. Uneven growth of the potato plant, and therefore uneven growth of tubers, can result in abnormalities in tuber shape. Fluctuating temperatures, moisture, and nitrogen, especially in the formative stage, can cause malformed tubers. Uneven growth caused by variable climactic conditions is worsened by poor plant stand and variability between plants caused by variable seed-piece types, sizes and spacings.

Importance of Tuber Quality

Potatoes are a nonfattening, nutritious and wholesome food that supplies many important nutrients to the diet. Potatoes contain approximately 78% water, 22% dry matter (specific gravity) and less than 1% fat. About 82% of dry matter is carbohydrate, mainly starch, with some dietary fibre and small amounts of various simple sugars. Although potatoes contain only relatively little protein (0.6-1.2% of the fresh weight), their nutritional quality is better than cereals or soybeans. Potatoes contain at least 12 essential vitamins and minerals, are a source of vitamin C, thiamine, iron and folic acid.

Quality influences consumer preference and product saleability. Good quality potatoes are clean, uniform in shape and size, have an unmarked skin and firm flesh, are free of internal defects, and have shallow eyes. Consumers do not like the waste of trimming potatoes with deep eyes and surface defects. Other tuber defects that may adversely influence quality are greening, secondary growth, growth cracks, scab, storage rots, internal black spot, skinning, Rhizoctonia sp. sclerotia, bruising and skinning or other mechanical damage. In addition to physical quality and appearance, fresh market potatoes must also have good cooking and eating quality.

The main cooking quality factors for home-prepared potatoes are texture, colour, flavour and odour. The texture of a cooked potato is directly related to dry matter content. Potatoes are described as mealy and dry if they have a high dry matter content, or moist to wet with a low dry matter.

Potatoes such as Russet Burbank, which have a high dry matter content, have a dry, mealy texture when baked, but when boiled they may slough or break up. Potatoes, such as Norland, with a lower specific gravity are inclined to be undesirably wet when baked, but hold together well when boiled. White-fleshed potatoes are expected to have a creamy white colour when boiled or baked. Yellow-fleshed varieties should retain their characteristic yellow colour during cooking.

White or yellow-fleshed potatoes should exhibit minimal after-cooking darkening. This darkening, which is often more prevalent at the stem end of the tuber, is caused by an accumulation of a dark pigment created by the reaction between chlorogenic acid and iron, both naturally occurring compounds in potatoes. Varieties such as Norland, which naturally contain higher amounts of chlorogenic acid, darken more than varieties containing less chlorogenic acid. In the home, after-cooking darkening may be controlled with lemon juice or vinegar (which contain acid).

In the commercial potato processing industry sodium acid pyrophosphate is used to reduce darkening after cooking. Cooked potatoes are expected to have "normal" potato flavour and aroma, without off-flavours or odours. Glycoalkaloids, such as solanine and chaconine, which are present at low levels (2-15 mg/100g fresh tuber weight) in all potatoes, represent important flavour components. Collectively, this class of compounds is referred to as total glycoalkaloids (TGA). If TGA levels are between 10 and 20 mg/100 g fresh weight the potatoes are bitter when eaten and are potentially toxic if levels are greater than 20 mg/100 g.

Twenty (20)mg TGA/100g fresh weight is the maximum allowed by Health Canada. Surface greening of the tubers often signals the development of bitterness due to TGA accumulation, although the development of the bitter alkaloid compounds and green colour are independent. When potatoes are exposed to light, the green pigment chlorophyll forms, and independent chemical reactions in the outer layer of the potato tuber are triggered, increasing TGA levels. Moderately green

tubers should not be consumed without first taste testing the peel of several raw tubers. Bitterness or a burning sensation indicates that TGA levels are too high and the potatoes should be discarded.

Elevated TGA levels in potatoes may result not only from exposure to light but also from harvesting immature potatoes, bruising, skinning, improper storage temperature control, wet conditions prior to harvest, chilling or freezing. Potatoes harvested cold should be slowly warmed to 55-60°F (13-15°C) for 14 days after harvest before the temperature is reduced for long-term storage. Unlike other problems associated with potato quality such as after-cooking darkening, TGA accumulation cannot be counteracted with chemicals or special processing techniques. Not all glycoalkaloids leach from potatoes into blanching or cooking water, and they are not destroyed at the high temperatures associated with boiling, baking or deep-frying.

Quality of French Fry

Four primary factors determine French fry quality: dry matter, levels of reducing sugars, defects and flavour. One of the most important qualities of French fry potatoes is high dry matter, which is measured as specific gravity or percent solids. As the specific gravity increases, the water content of the potato decreases. French fry processing removes water from the potato, hence the higher the water content of the potato, the more water or weight lost during processing resulting in a lower yield of fries. Potatoes with a high specific gravity produce a greater yield of fries than potatoes with low specific gravity.

Specific gravity also has a direct bearing on the amount of oil absorbed by the potato slices during the deep frying process. Fries made from tubers with low specific gravity absorb more oil than fries from tubers with high specific gravity. Although some oil absorption during deepfrying is desirable for flavour development, too much results in limp, greasy fries. Low specific gravities increase production costs

because more oil is used in the frying process. French fries should have a uniform light cream to golden colour. French fry colour is largely determined by the reducing sugar content of the potato tuber: potatoes with high reducing sugar levels make dark fries. When potatoes are fried, the reducing sugars react with amino acids in the tuber to form dark products in a nonenzymatic browning reaction.

The concentration reducing sugars in the tubers depends on variety, growing conditions, maturity, and storage conditions. The processor can make some adjustments to fry colour during processing, primarily by manipulating blanching conditions. Plate #2 shows the range of fry colours that occur. Processors prefer the lightest fry colour shown in Plate 2. Defects such as sunburn, net necrosis, sugar ends, bruises, carbon spots and light and dark brown areas, negatively influence the appearance of the finished French fry and reduce its value on the market. High quality French fries have a good potato flavour, are free from rancidity, bitterness, or off odours.

Their external surfaces should be moderately crisp showing no separation from the inner portion. The inside is tender, mealy and free from sogginess. The contrast in texture between inner and outer portions contributes to the pleasurable textural sensation when the fries are chewed. Some fries have a coating applied to enhance this textural contrast.

Growing Potato Plants

Up to the 1800s, growing potatoes was only for the poor. Their nutritional value and taste has been appreciated more over the last 200 years and the humble potato has now become part of the standard diet of many countries. Potato growing takes up a large amount of your garden space, so think carefully before planting. It is best to concentrate on the early potato varieties and those maincrop varieties which are unsuitable for commercial growing.

A potato plant will grow in most soils as long as it has been well dug. The soil should be clear of large lumps and stones. Add as much compost as is available, or bonemeal (or similar) at the time of digging. It is a good idea to prepare the soil in mid November or December to allow it to settle before planting and let the compost begin to break down. Although not fussy about the soil type, the potato prefers a slightly acidic soil, so do not add lime before, during or after planting.

The ideal position for a potatoes is a warm and sunny site with little or no shade. Do not grow potatoes in the soil for two years in succession, this will increase the risk of disease. For the same reason do not plant potatoes where tomatoes were grown the previous year. Both are fromily and will readily transmit diseases to each other.

Planting Potato Seed

Potatoes are grown from 'seed' potatoes which grow better if they are allowed to 'sprout' before planting. First purchase the seed potatoes in mid January. These are available from seed catalogues or your local garden centre. Look for seed potato which is certified as free from disease and select healthy looking examples about the size of an egg. The disadvantage of the catalogues and garden centres is that the number of potato seed varieties is very limited and they are only sold in large amounts. If you only want more variety and to buy several varieties in small amounts, the solution is to go to one of the many potato shows around the country. For an idea of a typical show and prices etc.

Place the seeds in boxes (for small amounts, cardboard egg boxes are ideal) in a light airy position at a temperature of roughly 10^{0}C (5^{0}F).

The potato seed should be positioned so the sprouts are uppermost and the 'stalk' end (where they were severed form the parent plant) is at the bottom. Sometimes this is a bit difficult to judge, but if you get it wrong, and the potatoes sprout from the bottom end, simply rub off the sprouts and turn the potato to the correct position. The picture below

shows the stalk end of a potato which should be at the bottom when placed in the egg boxes. As the potato seed produce sprouts, remove all but the topmost four to ensure that they receive all the goodness of the seed potato. The ideal sprout length at planting time is 2.5cm (1in) although this is not critical. What is critical, is that the sprouts are green and not white coloured. White sprouts are caused by not enough light.

If the sprouts appear too early for planting the potatoes outside, simply rub them off cleanly and they will re-sprout in a couple of weeks time. Research has shown that the sprouts can be removed five times without any ill effect on the sprouts which will replace them. Early to mid March is the ideal time to plant the sprouted potatoes in the open ground. Plant 'earlies' about 30cm (12in) apart from each other, in rows which are 60 cm (2ft) apart. Plant maincrop potatoes about 35cm (15in) apart, in rows which are 75cm (2ft 6in) apart. Where you are planting more than one row, the rows should (ideally) run from North to South to allow each plant it's full share of sun.

Dig a trench about 10cm (4in) deep, placing the potatoes in it with the sprouts pointing upwards. Hand fill the trench over the potatoes trying to avoid damaging any sprouts. Scatter bonemeal or similar long lasting fertiliser over the top soil and rake it in.

Planting Potatoes For Christmas Eating

In many areas it's quite possible to grow your own potatoes for eating in October through to Christmas. Potato plants produce good sized tubers in 65 to 100 days depending on variety.

Frost damage is the first concern in the early stages. If shoots emerge above the soil level and frost threatens, draw a little soil from the bed edges over them. After the potatoes plants have grown to about 20cm (8in) pull up the soil in between the rows around the plants leaving a few centimetres still showing. Repeat the exercise in two to three weeks time. The diagram to the right shows the ideal end result. During

the growing season, ensure that weeds are removed and apply some long lasting fertiliser again around mid-August. A month or so after planting, the dense foliage of the plant should then block out sufficient light to deter all but the most vigorous weeds.

As the potato plant grows, do not remove or 'stop' the foliage, it is supplying food to the tubers. However, remove any flower heads or buds which appear because these will produce potato seeds which will divert energy from the potato tubers.

Potato Pest and Disease

Potatoes are ready for harvest when the foliage first starts to die and turn yellow. Early (new) potatoes can be lifted earlier (no earlier than June) to get the very tastiest potatoes. In this case, harvest them about a week after the potato plant flowers first appear. New potatoes only produce a couple of handfuls of potatoes per plant, so dig up the whole plant.

Maincrop potatoes will be ready for harvest in mid-Autumn. For all potatoes, dig them up from the side of the ridge to avoid damaging the potatoes. The foliage can be placed on the compost heap only if it completely free from disease—if not, burn it. If you don't need all the potatoes from a plant at one time or if you want a few early in the season, simply burrow around the roots with your hands and remove the potatoes you need. The remaining potatoes will continue to grow.

Remove any soil clinging to the potatoes and leave them on the soil for a few hours to dry out—if they are stored damp, they will rot. Leaving them on the soil surface in the sun will harden the skin slightly, doubling the storage time—this is especially important for maincrop potatoes. Store the potatoes in boxes or sacks, checking them every few days, removing all but those in good condition. Damaged or blemished potatoes should be eaten immediately. One word of caution, some potato flowers may well turn to berries which are definitely poisonous. Eaten in sufficient amounts the berries can be fatal.

Other Methods of Growing Potatoes

There are two common alternate methods of growing potatoes by the amateur gardener. The first is container growing. One seed potato should be placed 15cm (6in) deep in a container which is 30cm (1ft) or more in diameter. Normal potting compost can be used. The amount of potatoes will obviously be small, but it's fun for the small garden. The second method is to cover the planting area for early varieties with black polythene, simply held down with earth or bricks. The potatoes are planted at normal depth by cutting a '+' shaped hole in the plastic and planting with a trowel. The advantages are that the soil is warmed by the black polythene, giving the young plants protection from frost. The potatoes should be ready for harvest a couple of weeks earlier than normal.

Pests and Diseases

Potatoes grown from stock certified as free from disease and planted in soil not occupied by potatoes the previous season, are relatively free from pests and diseases. The most common ailments to affect potatoes are potato blight, scab, slugs, and wireworm. There is little protection available that can effectively solve these problems.

Potato blight has two phases—the first attacks the leaves, the second attacks the potatoes themselves. If your potatoes are attacked by the blight, their is no cure. However, cut the foliage back to 5cm (2in) from the ground at the first signs of infection. This will prevent the blight from getting to the potatoes themselves. With no foliage, the potatoes will not give of their best, but they will continue to grow for a month or so.

Potato Blight

The first signs of potato blight are dark brown markings on the edges of the potato plant leaves. The next stage is that the leaves curl and the whole potato plant wilts and eventually falls over. This happens extremely quickly and the whole process can take only a matter of days.

Treatment of Potato Blight

Once you have it there is no treatment. To help avoid potato blight there are two main courses of action:

- — Destroy all infected plants by burning them. Do not plant potatoes (or tomatoes) on the same patch of land without leaving an interval of three years.
- — Spray potato and tomato plants with Bordeaux Mixture (available at most garden centres) in May and June to help prevent infection.

As far as we are aware, eating potatoes infected with potato blight will not cause serious medical problems.

Crop Rotation

The fungus which causes potato blight goes under the name of Phytophthora infestans. It is spread by either wind or rain in temperatures above 10°C (50°F). Avoid it occurring in subsequent years by good crop rotation and burning all infected plants as soon as the disease is noticed.

References

Larry Zuckerman, *Potato, The: How the Humble Spud Rescued the Western World*, Douglas & McIntyre, 1999.

Lang, James, *Notes of a Potato Watcher*, Texas A&M University, College Station, TX, 2001.

Salaman, Redcliffe N., *The History and Social Influence of the Potato*, Cambridge University Press, 1989.

Hawkes, J.G., *The Potato: Evolution, Biodiversity & Genetic Resources*, Smithsonian Institution Press, Washington, D.C., 1990.

Stevenson, W.R., *et al.*, *Compendium of Potato Diseases*, 2nd ed, Amer. Phytopathological Society, St. Paul, MN., 2001.

3

Commercial Potato Production

Potatoes are a crop that can be grown and marketed quite successfully at the retail level, especially at local produce auctions, farm markets, and roadside stands. The diversity of potatoes is just beginning to be realised as more speciality potatoes with different skin or flesh colours are being grown and marketed. The use of different colours adds to the visual appeal of potatoes on display and can attract attention at a retail outlet.

Potatoes grown for the fresh market are marketed in Pennsylvania from mid-July through late September if not stored and from mid-September until mid-May if held in storage. Fresh-market potatoes are sold loose, in containers, and in 3 lb to 50 lb paper or poly bags. Five basic marketing alternatives are available to the potato grower: wholesale markets, producer cooperatives, local retailers (grocery stores), roadside stands, and restau- rants. In wholesale marketing, producers negotiate price with the retail chain stores. Cooperatives such as the Pennsylvania Potato Growers Cooperative market potatoes to a wide variety of outlets.

Local retailers are another possibility, but you must take the time to contact produce managers and provide high-quality potatoes when the stores require them. Roadside stands provide opportunities to receive higher than wholesale prices for your potatoes, but you may have some additional expenses for advertising, building and maintaining a facility, and providing service to your custom- ers. Some growers sell

to local restaurants. Processors are not likely to contract with small acreage growers.

Importance of Field Selection

Field selection is one of the most important contributors to high potato yield and quality. Potatoes are a high-value crop and should only be grown on the best land available. Several factors must be considered before a field is selected for potatoes. Some factors such as drainage and herbicide residues can have a significant impact on yield while other factors, such as soil texture and crop rotation, have a lesser impact.

Crop Rotation

Potatoes must be rotated with other crops to slow the accumulation of diseases and reduce the impact of insects such as Colorado potato and flea beetles. When planning crop rotation two factors must be considered:

- The length of rotation
- The rotational crop

The length of time between potato crops is determined by disease and insect levels. A two or three year rotation adequately controls most diseases as well as reduces the population of Colorado potato beetles and flea beetles. Other diseases such as Verticillium wilt, require longer rotations. A four-year rotation is strongly recommended. Shorter rotations result in a build-up of soil borne diseases and a condition called early dying. This condition can result in yield losses as high as 30%. Cereal grains and corn are generally the best to use in the three years between potato crops. However, the decision about which crop to use in rotation must be made considering:

- The similarity of diseases between potatoes and the rotational crop
- Soil conservation
- The herbicides used in the rotational crop

Weather conditions play an important part in disease development. Most diseases are more severe in wet seasons; however, a few types of root rots are most severe under dry conditions. Very favourable conditions for disease development may cause a disease that is a low risk to become a high risk. Conversely, high-risk diseases may not materialise if conditions are unfavourable for disease development. The disease problems occurring on at least the previous four crops should be taken into consideration when planning a crop rotation. Continuous planting to the same crop can eventually cause low risk disease problems to build up resulting in severe crop losses.

Influence of Soil Texture

Mineral particles in soil are grouped into sand (2 - 0.05 mm in diameter), silt (0.05 - 0.002 mm) and clay (less than 0.002 mm in diameter). The proportion of sand (S), silt (Si) and clay (C) particles present in a soil is referred to as texture. The presence of larger particles in soil is recognised as gravel, cobbles or stones. Soil texture strongly influences the infiltration of water, the ability of the soil to retain moisture (water holding capacity), its general level of fertility, the tendency to form clods and ease of cultivation.

The combination of similar textural classes forms textural groups. Well-drained soils with loamy sand to sandy loam textures are considered most suitable for potato production. These soils have an adequate capacity to retain water, provide sufficient aeration for root and tuber development and favourable conditions for planting and harvesting. Sound management practices are required to minimise the potential for wind erosion on these soils. Farmers are successfully producing potatoes on silt loam, sandy clay loam, silty clay loam and clay loam textural classes even though these soils are not considered ideal for potato production.

These finer texture soils are prone to water erosion in undulating landscapes, poor to fair internal drainage and soil clod formation if tilled when wet. A soil that contains a large

amount of clay becomes sticky when wet and lumpy when dry. If this soil is too dry at harvest, it is difficult to separate the soil clods (aggregates) from the potatoes. Tubers harvested from a wet clay soil will also require washing to remove soil particles. These soils are not considered suitable for commercial potato production.

Topography

Topography refers to the percent slope and the pattern or frequency of slopes in different directions in a field. The steeper the slope, the less suitable the field is for potato production. Surface runoff occurs when the amount of water from rainfall or an irrigation application exceeds the rate of infiltration, which is strongly influenced by soil texture. The risk of soil loss is proportional to the length of the slope and the severity of the grade. Medium and fine textured soils are most susceptible to water erosion in sloping landscapes.

Soil Salinisation

Salinity refers to the presence of water-soluble salts in soil. Saline refers to a soil condition where water-soluble salts are present in sufficient amounts to affect crop growth. Salinity is measured as the electrical conductivity (EC) and is expressed as deciSiemens/metre (dS/m) or microSiemens/ centimetre. The accepted standard is dS/m, however all units of measure are equal. Salinity levels range from less than 4 (non-saline), 4-8 (slightly saline), 8-15 moderately saline and >15 strongly saline. Soils with a reading of less than 4 are most suitable for potatoes. On moderately saline soils, growth and yield of potatoes can be reduced. Leaves may be darker and have burned edges. Plants growing in saline soil are unable to draw water and nutrients, particularly phosphorus, from the soil. Potato production plays a role in soil salinisation.

The addition of high mounts of irrigation water and fertiliser add significant quantities of salt to the soil. Early detection of salt-affected areas is critical in preventing crop damage or an increase in salinity. A field assessment determining the degree of salinity can be accomplished by soil

sampling and analysis or by mapping with an EM38 electromagnetic induction meter. If salinity is a concern, assessments should be conducted every 5 years to determine if salinity is increasing over time. If salinity is detected, growers should implement appropriate water management and cropping practice strategies to improve the productivity of the soil. Prairie Farm Rehabilitation Administration and private soil laboratories provides these service.

Soil Reaction

Soil reaction, or pH, refers to the degree of acidity or alkalinity. A pH between 7.1 and 14 indicates an alkaline condition, a higher pH value indicating a greater alkalinity. A pH less than 7.0 indicates acidity and the lower the pH the greater the acidity. Most nutrients are readily available in soils with a pH range of 6 to 7, with a decrease in availability at soil pH above or below this range. Strongly acid soils have low extractable calcium and magnesium, a high solubility of aluminium, iron, and boron, and a low solubility of molybdenum. At the other extreme are alkaline soils. Calcium, magnesium and molybdenum are abundant in these soils with little or no toxic aluminium. Phosphorous availability may be greatly reduced at both very low and very high pH values due to phosphate fixation.

Organic Matter

Organic matter is derived from the residues of decaying organisms such as plants, animals, insects and microorganisms. Organic matter is a very important component of soil that stores and supplies plant nutrients, holds soil particles together, improves tilth and reduces the risk of erosion. Organic matter increases soil porosity and promotes water infiltration. To maintain organic matter, the rate of addition of crop residues must equal the rate of decomposition.

Continuous production of row crops such as potatoes, beans or sugar beets usually results in a rapid decline of soil organic matter because the total amount of crop residue returned to the soil is low. The level of organic matter in the

soil can be maintained by crop rotations that include perennial forage crops, by adding manure, or by the production of cereal crops where residues are not removed and minimum tillage is practised.

Compaction

Heavy equipment traffic and excessive tillage cause soil compaction. Soil compaction can be identified by the following symptoms:

- Excessive clod formation
- Slow water infiltration, especially in wheel tracks
- Distorted root and tuber growth and premature wilting from shallow rooting

Potatoes are sensitive to the physical condition of the soil. Dense or compacted soil interferes with root penetration as well as water and nutrient uptake. The plant will display symptoms of stress and the yield can be reduced. Compacted soil zones can be identified by carefully inspecting root growth patterns and soil texture in a 3-foot (0.9 m) deep trench. Minimise compaction by reducing the amount of traffic or weight in the field. Avoid unnecessary vehicle traffic, tillage or hilling.

The effect of compaction can be reduced by the use of tillage equipment that loosens soil to approximately 14 inches (35 cm) without inversion of soil layers. Subsoil tillage is expensive because of the high capital and operating costs of the equipment. A potato grower should confirm that compaction is causing production problems before investing in sub-soiling equipment.

Drainage

Soil drainage refers to the ability of water to flow downward through soil. Well-drained soils have sufficient surface runoff and/or downward movement of water through the soil to result in relatively short periods of saturation. Poorly drained soils have a greater frequency and duration when the soils are saturated in all or part of the root zone. Soil drainage may be

influenced by topography, uniformity of soil materials, proximity to the water table and climate. Excessive water in the soil limits the free movement of the oxygen necessary for healthy root and tuber development. Excessive soil moisture increases the incidence of fungal diseases, and delays spring tillage, planting and harvesting. In extreme cases, excessive soil moisture causes tuber rot leading to the loss of the entire crop. Improved surface and subsurface drainage can reduce the effect of excessive moisture conditions in imperfectly and poorly drained soils.

Stoniness

Soils that are free of stones or coarse fragments are most suitable for potato production. Factors such as size and abundance of stones must be considered when assessing a field for potato production. Large stones and boulders in a field make tillage difficult and can damage harvest equipment. Stones increase the chance of potato bruising during harvest. Stones cause a problem in seed production if they get into the cutter.

Herbicide Residues and Re-cropping to Potatoes

Herbicide residues from previous crops grown in rotation with potatoes can affect yield and tuber quality factors such as size, shape and specific gravity. The vigour of seed produced from fields with an herbicide residue can also be negatively impacted. The herbicide residue is concentrated in the tuber resulting in poor seed performance. All farmers should keep detailed records of herbicide use for each field. When leasing or purchasing new land, review the field history for potential herbicide residues. Cultural and environmental factors, as well as the properties of the herbicide, influence the persistence of herbicide residues in the soil. These factors are highly variable between fields and growing seasons. Soil texture and moisture are the two most important factors affecting the re-cropping interval.

Microbial Activity

Soils with a higher organic matter and clay content tend to have a higher moisture holding capacity. Generally, the population of soil microbial flora and fauna is larger and more active in soils with a higher soil moisture level, organic content and temperature. The breakdown of herbicide residues are accelerated in situations where microbial population flourish.

Precipitation and Irrigation

The amount and distribution of moisturereceived during the growing seasons between the herbicide application and re-cropping to potatoes is the most important factor determining the rate of residue degradation. Degradation is accelerated in soils with high moisture content. Higher precipitation and/ or irrigation amounts increase available soil moisture, which in turn increases soil microbial activity and the rate of herbicide breakdown.

Farmers should be very cautious if the growing seasons following the application of an herbicide known to cause injury received:

— below average precipitation, or

— average to above average precipitation, but poor seasonal distribution causing short periods of drought.

Either of these situations will prolong the persistence of herbicide residues.

Chemical Application

The timing of chemical application, the rate at which the chemical was applied, whether any overlaps occurred during spraying and whether chemicals from the same family group were applied in consecutive years are all important factors which will influence residue persistence. Applying herbicides later in the growing season (mid - late June) will increase the risk of herbicide damage. When an herbicide is applied late in the season, soil microbes have less time for chemical degradation before freeze-up, particularly under rain-fed crop production.

Herbicide labels often recommend different application rates to control different weed species or levels of infestation. Applying at the higher rate will lengthen the safe re-cropping interval. Overlaps of herbicide applications at the end of the sprayer boom or at headlands and "spot" spraying of weed patches with higher than recommended label rates will increase herbicide persistence. This results in a greater risk when re-cropping to potatoes. Applying herbicides from the same chemical family in consecutive years also increase the risk of residue damage when re-cropping to potatoes.

Re-cropping Interval refers to the length of time between the application of the herbicide and re-cropping to potatoes. One year after application refers to one cropping year. For example if an herbicide was applied in 2000 and the recommendation is to not re-crop to potatoes for one year after application then do not re-crop until 2002. A field bioassay is a test strip of potatoes grown to maturity in a field to determine whether herbicide residues will affect a crop. For example, 10 months after a particular herbicide was applied, a test strip of potatoes could be grown to determine whether potato yield and/or grade will be affected 22 months after initial herbicide application.

Select an area of the field that mostly closely approximates the soil conditions of the entire field. Test strips should be planted perpendicular to the direction in which the site was sprayed using standard planting equipment and following normal cultivation practices. Test strips should be of sufficient length to transect several spray swaths and should be at least one planter width wide. Insects and foliar diseases, such as late blight, must be controlled in order to accurately assess the affect of residues on yield and grade.

Reliability of the bioassay results will increase with the number and length of test strips planted. The test strips should be examined for possible stand reductions and other symptoms of injury such as reductions in plant vigour, yield and grade. If any injury, stand reduction or yield reduction occurs, do not plant potatoes until another bioassay is conducted the following growing season. It should be noted

that yield losses within a test strip might not be measurable unless the yield can be compared to an untreated area adjacent to the treated test strip.

A laboratory bioassay is the growing of plants in a greenhouse or growth chamber to determine if herbicide injury will occur. Soil samples are collected from a field suspected of having an herbicide residue and submitted to a laboratory for a bioassay test. The lab will grow a species of plant known to be sensitive to a specific herbicide or class of herbicides in the sample soil, along with the crop to be planted. If injury occurs to the test bioassay plants, then the potential exists for significant field crop injury.

Preparation of Field

Heavy equipment traffic and excessive tillage cause soil compaction, which reduces potato yield and quality. Minimise compaction by reducing the amount of traffic or weight in the field. Do not carry out unnecessarily tillage operations. Potato producers use a wide range of tillage equipment and techniques to prepare the potato seed bed. The variation in tillage practices results from the wide range of soil types used in potato production and the variety of tillage implements available for soil preparation. Regardless of the tillage system used, it must meet the following criteria:

— Incorporate trash - A balance must be maintained between incorporating an adequate amount of crop trash to facilitate planting while maintaining sufficient soil trash cover to prevent wind erosion. Crop residue must be incorporated into the soil to allow for trouble free planting. The amount of tillage required to do this will depend upon the soil type, the type of crop preceding potatoes and the ability of the planter to perform properly in crop residues. Tillage should not incorporate so much crop residue that the soil is susceptible to wind erosion.

— Produce good tilth without drying out the soil or producing soil clods - Tillage should produce enough

loose soil to allow the planter shoe to penetrate to the desired depth and to provide the hiller discs with enough loose soil to construct a proper hill over the seed. Tillage that dries out the soil surface will reduce emergence and vigour. Tillage of finer texture soil types (loams) with the wrong implement or at the wrong soil moisture content will produce clods that remain intact throughout the growing season and into harvest. Soil clods are difficult to separate from the potatoes on the harvester thus reducing efficiency and increasing cost. Hard dry clods that come into contact with the tubers on the harvester and other handling systems will cause blackspot bruising.

— Herbicide incorporation - Sufficient tillage is required to properly incorporate pre-emergence herbicides. Follow the chemical manufacturers' instructions regarding the tillage method required for herbicide incorporation.

Field preparation should be performed in as few operations as possible. Excessive tillage will increase the cost of production, the likelihood of wind erosion and the amount of soil compaction.

Conventional

Operations such as deep tillage or discing are performed in the fall and cultivation is performed prior to planting in the spring.

Advantages:

— Inexpensive - Low capital and operating costs

— High capacity - 15-25 acres/hour (6-10 ha/hr) depending upon tractor speed and width of implement

— Preserves surface trash depending upon number of operations and shovel design

Disadvantages:

— Can produce soil clods, depending upon soil conditions at the time of tillage

Effect of Rotary Power Cultivator

In Manitoba and Saskatchewan, producers growing potatoes on loam and clay loam soils use conventional tillage in the fall. In the spring, a single pass with a rotary power cultivator prepares the soil before planting. The rotary cultivator is a power cultivator that stirs the soil laterally rather than mixing the soil vertically like a rotovator.

Advantages:

- Clod free soil bed
- Preserves surface trash
- Preserves moisture, especially important in rain-fed production

Disadvantages

- Low capacity - 5-7 acres/hour (2-2.8 ha/hr) with a 20-foot (6 m) implement depending upon soil conditions and texture
- Costly - High capital and operating expenses
- Poor weed control - has a tendency to transplant rather than destroy weeds

Fall Bedding and Reservoir Tillage

A fall bedding and reservoir tillage (diking) implement performs both the tillage and hilling operations. In the fall prior to spring planting of potatoes, the field is tilled with a deep tiller or a double disc. Next, deep subsoil tillage is performed, hills are constructed and reservoir dikes or depressions are formed between the rows. The reservoirs capture water from melting snow and increase soil moisture.

In spite of the aggressive fall tillage, surface trash cover is preserved, reducing the potential for wind erosion. In the spring, the hills are packed with rolling wire baskets to break up soil clods that survive the freezing and thawing cycles over the winter, the hills are reshaped and the reservoirs are removed. The potatoes are planted directly into the preformed

hills. After planting, the final hilling is carried out and new reservoirs are formed between the rows.

Advantages:

- High Capacity - 8-11 ac/hr (3-4.5 ha/hr) in fall, 15 ac/hr (6 ha/hr) in spring
- Hills warm up faster in the spring
- Increased water retention
- Reduces runoff and drown-out in low areas as a result of precipitation or irrigation
- More even moisture between high and low areas of the field
- Clod free soil preparation
- Preserves soil trash

Disadvantages:

- Costly - high capital cost
- Requires a large tractor (>300 h.p.) in the fall

Conservation of Soil

Soil Erosion

Wind and water erosion resulting in serious soil losses can be a problem on the prairies due to a combination of climate, soil type, growth characteristics of the potato plant and management practices used in potato production:

- Most potato producing areas, especially southern Alberta, experience high winds when the fields are exposed and vulnerable to erosion.
- Potatoes are usually grown on sand or sandy loam soils, which are susceptible to wind erosion.
- The risk of wind erosion is greater when planting light textured soils to potatoes rather than cereals. Potatoes do not emerge from the soil for 3 weeks after planting and it takes an additional 2 - 3 weeks before there is sufficient vegetative growth to prevent erosion.

— Tillage, crop rotation and harvest practices used in potato production can leave fields without adequate crop trash cover for many months of the year. In the fall and spring prior to potatoes, some crop residue must be incorporated into the soil to allow for trouble free planting. Often there is too little trash cover remaining to prevent erosion. The potato harvest is similar to an aggressive tillage operation, which disturbs the soil, leaving little crop residue cover to prevent erosion.

Soil formation is a natural and ongoing process, but if wind erosion is visible, soil losses are greater than the rate of soil formation. One storm that removes a 1/4 inch (7 mm) layer of soil results in the loss of 44 tons per acre (40 tonnes per ha) of top soil and will require at least 10 years to be replenished; assuming no further erosion occurs. The loss of topsoil through wind erosion reduces yield potential by negatively affecting fertility, physical structure, tilth, permeability and water holding capacity. In addition, eroded soil from agricultural land can fill ditches reducing the effectiveness of drainage systems. In some jurisdictions, like Alberta, a fine can be imposed if a farmer fails to control erosion, and the expense of cleaning out the drainage or irrigation ditches is the responsibility of the landowner.

The erosion prevention practices listed below are discussed under three headings:

— Potato field preparation and planting

— Potato harvest and post-harvest

— Long-term strategies

Field Preparation

In the fall, the combine should be outfitted with a straw and chaff spreader to distribute residue evenly across the field. Additional flailing or harrowing may be necessary to help chop and distribute the crop residue. Even distribution of the straw allows the farmer to use fewer and less aggressive tillage operations in preparing the field for planting the following spring. Fewer and less aggressive tillage operations preserve

more of the trash cover reducing the potential for erosion. Tillage of light textured soils, where it is relatively easy to incorporate crop residues, should be delayed until spring. If fall tillage of a light-textured soil is carried out, it should result in no less than 65% of the ground covered with crop residue.

In Alberta, fall bedding and reservoir tillage are commonly used to prepare fields for planting to potatoes the following spring. The equipment used to form the beds preserves sufficient trash cover to limit erosion. Orienting the beds at right angles to the prevailing winds reduces the impact of wind on soil erosion. The construction of reservoirs between beds prevents soil loss from water erosion caused by snowmelt and heavy spring rains. In the spring, a minimum amount of tillage should be used in preparing the soil so that trash cover is maintained. The producer should choose tillage methods that incorporate just enough crop residues to allow for trouble free planting, but preserve sufficient cover to prevent erosion.

Where practical plant potatoes rows at right angles to the prevailing winds to reduce drifting in the spring. In Alberta, rows may be oriented north/south as a method of frost protection. In the north/south orientation, both sides of the hill receive equal amounts of sunlight maximising the solar radiation captured during the day. Strip cropping (alternating strips of potatoes with a crop less prone to erosion) can be planted to reduce the unsheltered width in erosion-susceptible fields. Crop strips should be perpendicular to the prevailing winds in May. The width of strips will depend on the size of equipment used in cropping and the susceptibility of the soil to wind erosion. Recommended strip widths vary from 76 feet (25 metres) in loamy sands to 305 feet (100 metres) in silty clay loams.

Annual crops such as corn and sunflowers can be used as wind barriers to protect the soil after harvest. Properly spaced annual barriers also trap and evenly distribute snow for effective winter cover. Annual barriers should be planted closer together than tree shelterbelts because they offer less protection.

Potato Harvest

In potato production areas where there are no stones, the potato harvester should be equipped with a vine shredder, which eliminates the need for tillage to anchor potato vines. Potato vines that are not anchored in the soil can blow off the field and accumulate in ditches, fences and shelterbelts. Shredding provides even distribution of chopped vine residue improving ground cover and reducing erosion potential. If the harvester is not equipped with a vine chopper, a tandem disc or a chisel cultivator equipped with coulters should be used to anchor, but not completely incorporate the vines. Seeding an annual cereal after the potato harvest will provide good protection from wind erosion until a new crop is established the next season.

The success of this method is dependant upon the harvest date and the growing conditions after harvest. The chances of establishing a cover crop after September 15th to 20th are poor. The fall cover crop can be seeded by air just prior to harvest or with a conventional seeder just after harvest.

The emergence is less even when seeding by air. In Alberta, irrigation may be required to obtain even emergence of the cover crop. Some farmers choose to use wheat, oats or barley as a cover crop because the crop will winter kill and not interfere with future cropping plans. When choosing an annual cover crop, keep in mind that you will want a crop that germinates and establishes quickly.

The malting varieties of barley are well suited to quick establishment. Many farmers will use a higher seeding rate to ensure a good ground cover in the field, i.e. barley at 1.5 bu/acre (62 kg/ha). Others choose to seed fall rye or winter wheat as a cover crop. A non-selective herbicide is applied the next spring to control the rye or wheat, or the crop can be lft to mature. Manure applied to the soil surface at 30-40 tons/ac (70-90 tonnes/ha) will reduce drifting. Researchers have demonstrated a link between high raw manure applications and an increase in the incidence of common scab, but this affect is rare. The nutrient content of the manure should be

taken into account when planning commercial fertiliser applications the following season.

Mulch can provide the same erosion control benefits as a well-established cover crop or well managed crop stubble. Broadcasting cereal straw with a round bale shredder onto the surface of erosion susceptible areas of the field will reduce wind erosion and add organic matter. The application of 1 to 3 tons/acre (2 to 6 tonnes/ha) will provide 65% ground cover, which is sufficient to prevent erosion.

Long-term Erosion Control

Fields susceptible to wind erosion can benefit from a systematic planting of shelterbelts. Trees should be planted perpendicular to the prevailing winds that occur in May. The prevailing winds at this time tend to have the highest velocity and fields have the least protective cover. A distance of approximately 660 feet (200 metres) between tree rows is recommended. The distance between green ash trees planted in the sheltebelt should be 6 to 8 feet (2.0 to 2.4 m) apart.

In Alberta, there is a growing trend to reduce a soil's susceptibility to wind erosion through the build up of organic matter in the soil. The addition of composted manure on an annual basis at rates of 2.5 to 5.5 tons/ac (7 to 12 tonnes/ha) is a safe method of amending the organic matter and reducing the soil's susceptibility to erosion over time. Planting cereal and forage crops, which produce large amounts of organic matter, in rotation with potatoes helps reduce soil erosion. Care must be taken when including bean and canola crops in rotation with potatoes, as neither provides much crop residue. For assistance with soil erosion problems, contact your provincial soil and conservation specialists or Prairie Farm Rehabilitation Administration (PFRA).

Fertility and Fertilisers

Potatoes managed for maximum productivity exert a heavy demand on soil fertility. Significant quantities of nutrients are

removed from the field in the harvested tubers, while additional nutrients are lost through erosion and leaching.

Nitrogen Nutrition

Of all the essential nutrients, nitrogen most often limits potato production. Consequences and symptoms of nitrogen deficiency in the crop are:

- — Stunted growth
- — Premature death of vines
- — Poor yields
- — Increased susceptibility to diseases such as early blight and Verticillium wilt
- — Likewise there are consequences of excess nitrogen:
- — Negative impact on environmental - particularly due to nitrate leaching below the root zone
- — Excessive vine growth, which can interfere with harvest
- — Delayed tuber set leading to poor yields in regions with a limited growing season.
- — More deformed tubers and hollow heart
- — Tubers are physiologically immature at harvest, resulting in more harvest damage, poor storability and processing quality
- — Low specific gravity

Attention should be paid to nitrogen management because of its influence on crop productivity, quality and the environment. Nitrogen management refers to placement methods, timing of applications and assessment of N status as well as N application rates.

Nitrogen Placement and Time of Application

The relative efficiency of any nitrogen application varies depending on soil moisture, soil temperature, soil type, weed and crop growth. In Table 1 average efficiency of times and method of N application are compared to spring broadcast, which is given a value of 100.

Table 1. Average efficiency of times and method of N application are *compared to spring broadcast*

Time and method	Relative efficiency value
Spring broadcast	100
Spring banded	120
Fall broadcast	80
Fall banded	100

Efficiency values are based on nitrogen uptake by plants. Broadcast values assume urea-based nitrogen fertilisers are incorporated to minimise losses through volatilisation of ammonia. This incorporation step is critical on soils with a high pH or which contain free lie near the surface, particularly when rainfall is not received following application. Banding refers to applying fertiliser in-furrow or in bands below and to the side of the seed piece. Banding the N fertiliser restricts contact between the fertiliser and the soil. This delays nitrification of ammonia forms of N fertiliser, thereby reducing losses of nitrate-N. Banding also reduces loss of N due to contact with crop residues or immobilisation.

Excessive rates of banded N can impair sprouting and reduce plant emergence. In-furrow applications of N while seeding, should not exceed 50 lbs N/ac (45 kg N/ha) in the urea form, while no more than 80 lbs N/ac (70 kg N/ha) nitrogen in urea form should be applied as a sideband application. Fertiliser bands established prior to seeding should not be disturbed prior to or during the seeding operation to optimise N efficiency. Nitrogen losses due to leaching, denitrification, immobilisation and weed growth are generally higher for fall-applied than for spring-applied N.

These losses are further increased if the N is applied too early in the fall (prior to mid-September) or when soil temperature at the 4" (10 cm) depth is greater than 41°F (5°C). For these reasons, nitrogen should be applied in spring or in-season. In dry soil, banding N in late fall is as effective as spring banding. Nitrogen is relatively mobile in moist soil and therefore need not be applied near the plant to be effectively

used. However, surface application or shallow incorporation can result in reduced uptake if the seedbed remains dry during the early growing season. Nitrogen applied in a band is more effective than surface applications when the soil is dry.

Split applications for nitrogen

Splitting the total N application is an effective strategy to increase fertiliser use efficiency while limiting nitrate leaching. Part of the crop's total N requirement is supplied prior to planting or is banded at planting. This initial application of N is supplemented with one or more applications of N. The first increment of N is usually applied after tuber initiation. Additional later applications may be used if a petiole nitrate analysis indicates a N deficiency.

Several methods are available for applying N after crop emergence.

— The most common method is top-dressing which involves broadcasting granular fertiliser over the growing crop. Ammonium nitrate is the preferred nitrogen source for top-dressing because it does not require incorporation after application. Urea can be top-dressed, but hilling or irrigation must be used to incorporate the fertiliser shortly after application. Liquid N can be dribbled onto the hill at the base of the plants, but it also must be incorporated after application. Care must be taken to avoid contact between the foliage and liquid N, as it will burn the leaves.

— Nitrogen fertilisers may be banded in the soil to the side of the emerged crop row. Delaying banding increases pruning of the roots and may reduce yields.

— Fertigation, which involves the injection of liquid fertiliser into the irrigation water, can be used to apply N and other nutrients throughout the growing season. Liquid urea-ammonium nitrate (28%) is the most common source of N injected into irrigation water. Uniformity of N application depends on the uniformity of water application from the irrigation system. Centre

pivot and linear irrigation systems provide the most uniform application of water. Travelling guns and side-wheel roll systems do not distribute water as uniformly over the field as centre pivots or linear systems because of overlaps between moves. These systems should only be used to apply N fertiliser when the wind is very low and no other N application method is available.

Nitrogen assessment

Soil testing to a depth of 24 inches in the spring is critical to planning an effective and efficient N management programme. Post harvest soil samples may help growers to select succeeding crops, which will make maximum use of the residual N after the potato crop. Commercial soil test laboratories will provide N fertiliser recommendations based upon the nitrate soil test. The recommendations may vary with the soil type, the cultivar, end use of the crop, yield goal, the previous cropping and manuring history of the field. Each provincial soil specialist can also provide rate recommendations based on the soil test results.

Nitrogen management based solely on a soil nitrate test is challenging. In intensively cultivated and irrigated potato fields, mineralisation of organic matter may add significant amounts of N. Even in relatively fertile soils, leaching or very high plant demand can cause in-season N deficiencies. The nitrogen demand by the crop during tuber bulking may be 2.5 to 3.5 lbs N/ac/day (2.2 to 3.0 kg N/ha/day).

Petiole nitrate sampling allows for in-season monitoring of the crop's nutrient status. Collect the 4th petiole from 30 - 50 randomly selected plants throughout the field. Strip the leaflets from the petiole and keep the sample cool until shipping. Tissue samples are often collected weekly to track changes in nitrate levels, and to plan supplemental fertiliser applications should levels drop below optimum. Producers should follow laboratory instructions on tissue sampling techniques and interpretation of petiole nitrate results.

Critical petiole nitrate-levels decline as the potato crop develops and matures. Generally, petiole nitrate-N levels at

tuber bulking are <10,000 ppm = low, 10,000-15,000 ppm = medium, >15,000 ppm = sufficient.

Phosphorus

Soil Availability and Plant Uptake

Soil phosphorus (P) occurs in many forms depending on the pH and chemical composition of the soil. Only a small fraction of the total amount of P present in the soil is actually available to the crop within a given growing season, but the supply of available P is constantly replenished from soil reservoirs of less available P. Consequently, soil P levels are usually fairly stable over the course of the growing season and do not change much from year to year unless the field has been heavily fertilised. The roots absorb phosphorus ions dissolved in the soil water.

Phosphorus deficiencies can occur even in soils with abundant available P if drought, low temperatures or disease interfere with the phosphorus diffusion to the root through the soil solution or otherwise impair development and function of the roots. Potato plants require an adequate supply of P throughout the growing season to achieve optimum yields. During the early growth stages, P stimulates the development of a vigorous root system and healthy tops. Plant demand for P peaks at tuber set and then slows during bulking as the nutrient needs of the developing tubers are met by translocating P from the tops of the plants.

Phosphorus deficiencies slow growth and reduce yields without producing any obvious changes in the appearance of the crop. Only after P deficiencies have become severe will the crop begin to develop visible signs of P stress - stunted dark green leaves and the edges of the youngest leaves begin to fold downwards. As P deficiencies are difficult to diagnose based on the appearance of the crop, P status should be monitored by tissue testing. Optimum tissue P levels vary from 0.2-0.5%, depending on the crop growth stage.

P Fertilising Strategies

The P requirements of potatoes are much higher than the P

levels typical of most Prairie soils and some type of P fertiliser programme is usually required. Phosphorus application programmes should be based on the recommendations obtained from fall or spring soil tests. As P fertilisers are relatively expensive and the nutrient is immobile in the soil, the objective of the ideal P fertility programme is to raise levels of available soil P in the effective root zone to precisely match the requirements of the crop. Band application of P fertiliser below and to the side of the seed row represents the most efficient means of meeting crop P needs with the minimum amount of P fertiliser. Placing the P fertiliser in a concentrated band adjacent to the developing root system increases the efficiency of fertiliser recovery by the roots while also reducing fixation of the applied P into forms unavailable to the crop.

Banding is particularly effective in soils testing low in P or with high P fixing capacity and in situations where growing conditions may interfere with development of a healthy root system. Broadcast/incorporating P fertilisers prior to planting represents a convenient option in soils requiring only limited P amendment. Although P is relatively immobile in soil, timely application of P as a sidedress treatment followed by irrigation or via fertigation with P may be useful if tissue tests indicate the crop is at risk of a P deficiency. This practice is only recommended when there is risk of a yield limiting nutrient deficiency and there are no other application options available. Excess application of P as either fertiliser or manure is economically and environmentally unsound but has little potential to reduce yield or quality.

Potassium

Soil availability and plant uptake

Potassium (K) is supplied to the crop root through diffusion over short distances; hence K is also referred to as a relatively immobile nutrient in soils. Much of the K available to the plant is held on exchange sites on clay or organic matter, which replenishes the soil solution as rcot uptake of K occurs. Consequently, soil K levels are usually stable over the course

of the growing season and from year to year. Potatoes require abundant soil K, as this nutrient is crucial to metabolic functions such as the movement of sugars from the leaves to the tubers and the subsequent transformation of those sugars into starch. Potassium deficiencies reduce both the yields and quality of the potato crop. Potassium deficiencies impair the crop's resistance to diseases and its ability to tolerate stresses such as drought and frost. Tubers adequately supplied with K are more resistant to blackspot bruising or after cooking discoloration, while also experiencing less moisture loss and disease during storage.

Potassium deficiencies typically first show up in the sandier or knoll areas of the field. As K is mobile within the plant, the lower leaves are first to display symptoms of K deficiency - yellowing of the leaves with scorching around the leaf margins. Severely deficient plants may take on a coppery sheen. Small dead spots resembling the lesions produced by fungal disease may form on the leaves. As visible signs of K deficiency only become apparent once the crop is severely stressed, the K status should be monitored by tissue testing. Optimum tissue K levels vary from 2-5 %, depending on the crop growth stage.

K Fertilising strategies

Potassium application programmes should be based on the recommendations obtained from fall or spring soil tests. As K fertilisers are relatively inexpensive, many growers opt to add a small amount of K fertiliser to potatoes even in K rich soils. Potassium fertiliser applied in a band adjacent to the seed row may provide a jump-start to the emerging potato plants. To avoid salt damage to the crop, the total amount of all actual fertiliser nutrients applied as a band should not exceed 100 lbs /ac (88 lbs/ha). Broadcast/incorporating K fertilisers prior to planting represents the preferred approach in soils requiring more substantial K amendment.

Although K is only moderately mobile in the soil, timely application of KNO_3 as a sidedress treatment followed by irrigation or via fertigation may be useful if tissue tests

indicate the crop is at risk of a K deficiency. This practice is only recommended when there is risk of a yield limiting nutrient deficiency and there are no other application options available. Muriate of potash (KCl 0-0-60) is the least expensive and most readily available form of K fertiliser. There is some evidence that excessive Cl associated with heavy application of KCl may reduce the specific gravity of the potato crop, but this is rarely an issue on the relatively K-rich Prairie soils.

Micronutrients

Although micronutrients (boron, chlorine, copper, iron, manganese molybdenum and zinc) are only required in relatively small quantities by the potato crop, they are no less important to the health and productivity of the crop than are the other nutrients. Although deficiencies of micronutrients are relatively uncommon in Prairie soils, growers should still have the micronutrient levels in their soil evaluated prior to planting and should confirm the micronutrient status of their crop through tissue testing.

The micronutrient needs of the crop may be met through either soil or foliar applications. As micronutrients are relatively expensive and the margins between adequate and excessive supplies are often narrow, growers should exercise caution when utilising these products. Check strips represent a useful tool for confirming the benefits and cost efficiency of any fertiliser treatment.

References

Brook, R. C., R. J. Fick, and T. D. Forbush, *Potato StorageDesign and Management*, East Lansing, MI.: Michigan StateUniversity Agricultural Engineering Information Series, 1995.

Christ, B. J., *Potato Diseases in Pennsylvania*, PennsylvaniaCollege of Agricultural Sciences, University Park, PA. 1998.

Dunn, J. W., J. K. Harper, and G. L. Greaser, *AgriculturalAlternatives: Fruit and Vegetable Marketing for Small-scaleand Part-time Growers*, University Park, PA: Penn StateCooperative Extension, 2000.

Greaser, G. L. and J. K. Harper, *Agricultural Alternatives: Enterprise Budget Analysis*, University Park, PA: Penn StateCooperative Extension, 1994.

Lamont, W. J., *et al., Agricultural Alternatives: Irrigation for Fruit and VegetableProduction*, University Park, PA: Penn State CooperativeExtension, 2001.

Rowe, Randall, editor, *Potato Health Management*, St. Paul,MN: APS Press, 1993.

Sieczka, J. B. and R. E. Thornton, *Commercial PotatoProduction in North America,* Orono, ME: The PotatoAssociation of America Handbook, 1992.

Weisz, R., Z. Smilowitz, M. Saunders, and B. Christ, *Integrated Pest Management for Potatoes,* University Park,PA: Penn State Cooperative Extension, 1995.

4

Soil Management

The potential for erosion on soils under potato production is much greater than it is on soils under cereal and oilseed production. Potatoes are traditionally grown on sandy textured soils. They require at least three weeks to emerge, and another three weeks for enough foliage growth to provide adequate ground cover. After harvest, the potato fields have little crop residue remaining on the surface because very little plant residue is produced and much of it is lost during harvest. Between harvest and the establishment of a crop canopy the next growing sea son can be a period when fields are extremely susceptible to erosion.

Most of this erosion will be by wind. The forces of wind on an unprotected field will cause the fine and medium sand particles from 0.1 to 0.5 mm in size to move in a bouncing or jumping action called saltation. These particles are lifted by the windstrearn and then blasted onto the ground downwind. On contact, these particles dislodge other particles and force them into the windstream (Figure 1). This causes a chain reaction which increases in intensity as it moves across the unprotected field. Larger soil particles are dislodged and eroded by a process called surface creep.

Soil Management Practices

Farmers growing potatoes must be careful to use and manage sandy soils to maintain productivity both today and in the

long term. The following are some management options which are effective in reducing wind erosion on potato lands.

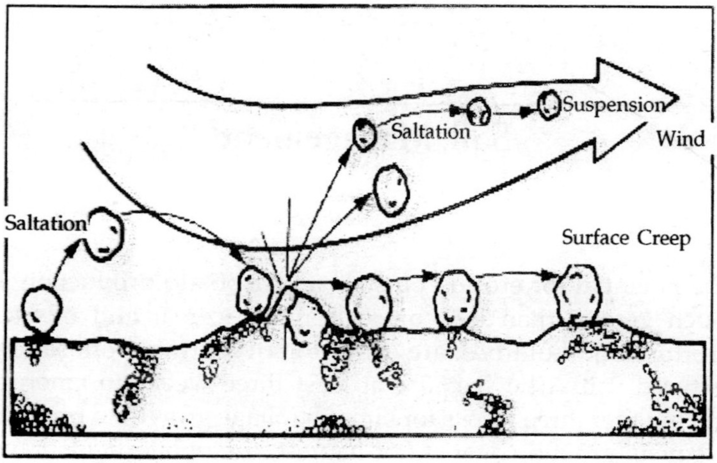

Figure 1. Surface Creep

Fall Seeded Cover Crops

Cover crops provide vegetative cover in fall I and early spring for soils susceptible to erosion. In a rotation which includes potatoes, small grains such as wheat, barley and rye can be seeded immediately after the potato harvest in July, August and the early part of September. Seeding small grains after September 15 will be beneficial only when the fall is long and open without a killing frost. The recommended seeding rate for small grains is a bushel per acre (65 kg/ha). To ensure that protection remains adequate, lands seeded to cover crops should not be tilled until the following spring.

Shelterbelts Planting

Fields in special crop rotations can benefit from shelterbelt planting. This is especially true of rotations which include low

residue producing crops. A distance of approximately 660 feet (201 m) between belts is recommended. Spacing between trees in the belt should be 6 to 8 feet (1.8 to 2.4 metres). Where row crops are included in the rotation, shelterbelts must be spaced wide enough to accommodate row crop equipment.

Strip Cropping

Strip cropping will reduce the unsheltered width of erosion prone fields and limit saltation. Narrower fields perpendicular to the wind reduce erosion because the distance the wind is allowed to travel across a field to start soil movement is decreased. The width of strips will depend on the size of equipment used in the crop rotation and the erodibility of the soil. Fields should be perpendicular to the May prevailing winds. The soil is most vulnerable to erosion in May, because the wind forces are strongest and fields have the least vegetative cover.

Crop Residue Management

Maintaining enough crop residue on the soil surface will prevent wind erosion. If the crop before potatoes is a small grain, the following practices are recommended:

— The combine should have a straw and chaff spreader to distribute residue material in the field evenly. Harrows may also be necessary to help spread the straw and chaff.

— Fall tillage operations should result in a 65 per cent ground cover on sandy soils before seeding. The number of tillage operations that result in this amount of ground cover for various wheat and barley yields are shown in Table 1.

— Plant directly into previously fall-tilled fields with a conventional or modified potato planter.

— If no fall or spring tillage is carried out, the conventional planter can be modified to plant directly into standing stubble.

The following practices are recommended in handling potato residue.

— Attaching a vine chopper to a potato harvester eliminates the need for discing to anchor vines in the fall. The distribution of chopped residue will improve ground cover after harvest. Also, the chopped vines will not roll up in the wind.

— Anchoring the vines with a disc type implement with two-thirds of the discs removed. This practice leaves some vegetative cover and increases the surface roughness (ridging) of the soil. If the discs are not removed, the implement should be set to anchor the vines into the soil rather than tilling the soil and burying the residue.

References

Boyle, M., W.T. Frankenberger, Jr., and L.H. Stolzy, "The influence of organic matter on soil aggregation and water infiltration", *Journal of Production Agriculture*, 1989.

Pimentel, D. *et al.*, "Environmental and economic costs of soil erosion and conservation benefits", *Science*, 1995.

Soil Association *Soil Management on Organic Farms*, Available from: Soil Association, 2003.

US Department of Agriculture, *Soil Biodiversity*, Soil Quality Information Sheet, Soil Quality Resource Concerns, January 1998.

5

Cover Cropping in Potato Production

Broadly defined, a cover crop is any annual, biennial, or perennial plant grown as a monoculture or polyculture, to improve any number of conditions associated with sustainable agriculture. Cover crops are fundamental, sustainable tools used to manage soil fertility, soil quality, water, weeds, pests, diseases, and diversity and wildlife, in agroecosystems. Agroecosystems are ecological systems managed by humans across a range of intensities to produce food, feed, or fibber. To a large degree, humans shape the ecological structure and function of natural processes that occur in agroecosystems.

As agroecosystems often interact with neighbouring natural ecosystems in agricultural landscapes, cover crops that improve the sustainability of agroecosystem attributes may also indirectly improve qualities of neighbouring natural ecosystems. Farmers choose to grow specific cover crop types and to manage them in specific ways based on their own unique needs and goals. These needs and goals are influenced by biological, environmental, social, cultural, and economic factors of the food system within which farmers operate.

In potato farming, cover crops can be grown to:

— protect the soil from erosion;

— provide fresh organic matter that will improve soil tilth;

— prevent the leaching of nutrients;

— serve as sources of nitrogen; and

— help in the management of certain pest problems.

Cover crops can be used in the following ways in potato production:

— As green manures established the year before planting potatoes. Several studies report an increase in yield and quality of potatoes with a preceding cover crop due to improved nutrient cycling and soil tilth. This seems especially true in warm climates where the turnover of nutrients is rapid, but has also been reported in cold climates. It is estimated that the effect of a preceding cover crop is equivalent to the application of 10 to 90 t/ha of manure, depending on whether the green manure was fertilised or not.

— As catch crops after potato harvest. A catch crop is a crop grown in the same year as the main crop but after the main crop. The main role of cover crops in this case is to protect the soil from erosion and to prevent leaching of nutrients unused by the potatoes. It can be tilled in the fall or kept as an overwintering cover crop that will be tilled in the spring.

— As a full-season crop in rotation with potatoes. The aim of this practice is usually to increase soil organic matter levels and involves turning in mature plants rather than green ones. It can also be part of a non-chemical weed or pest control strategy.

The types of plants to be used as cover crops can be classified in five main categories:

— Legumes;

— Cereals and grasses;

— Crucifers;

— Other families of plants;

— Mixtures of the above.

Let's examine the advantages and disadvantages of these plants and see how they can be used profitably in potato production.

Legume Species

The legume family comprises fodder crops like clovers, vetches and alfalfa, as well as pulse crops like soybeans, fababeans and lupine. Their ability to fix nitrogen from the air with the help of specific bacteria gives them a special value as source of "free" nitrogen on the farm. Unfortunately, legumes usually require for their optimal growth a higher soil pH than that needed to produce scab free potatoes. The most commonly used legume species for cover cropping are red clover and hairy vetch.

Small red clover is a favourite for cereal underseeding because of its low seed cost and the fact that it does not grow as tall as medium red clover, therefore not hindering mechanical harvesting. However, clovers can host verticillium wilt. According to a Dutch study, white clover undersown in cereals can have a negative effect on the following potato crop yield. Hairy vetch is a fast growing winter annual legume but does not survive winter which makes it useful as a cover crop. Pulse legumes are mostly used as rotation crops but can also be used as cover crops.

Soybean is a very good green manure before potatoes especially in warm climates. It is more tolerant of soil acidity than most legumes and is good against scab. A disadvantage of soybean grown as a crop is that it produces little residue and so does not contributes to the maintenance of soil organic matter levels. Lupine is a favourite in Eastern European countries as a cover crop preceding potatoes. Its effect on yields typically lasts for two years in the rotation.

Legumes can be planted as full season cover crops with a cereal nurse crop or as a catch crop in the year before potatoes. The advantage of the cereal nurse crop approach is that a harvestable crop can be secured. The legume develops fully after the cereal harvest. On light soils, it is plowed in the following spring and serves as a slow release source of nitrogen for the potato crop while protecting the soil all winter. In a German study, clover undersown in cereals gave higher potato yields the following year than either rape,

sunflowers or legumes sown after cereal harvest. Because legumes develop rather slowly development, they may not provide a good cover if planted as catch crops after the cereal harvest.

As is the case for cereals, it is usually too late after potato harvest to plant a legume, unless it is an early harvest. Sowing a legume at the beginning of August does not leave much time for development and is risky in a dry year. Hairy vetch and crimson clover develop faster than many other legumes and are recommended as catch crops after early potatoes. Unfortunately, their seeds are expensive and they will fix as little as one fifth the nitrogen of a spring sowing.

Cereals and Grasses

Role of Rye

In several countries, rye has been the most popular cover or rotation crop for potatoes, especially for light soils. Unlike legumes, it does not need a high pH and is very competitive with weeds. It is one of the best plants to prevent nitrate leaching in the fall. Rye is also very tolerant of low winter temperatures. Unfortunately, rye is difficult to kill in the spring and it can be a host for nematodes that attack potatoes. As a cash crop, rye has a limited market in Eastern Canada, although it can be used as animal feed. Rye is also very susceptible to ergot.

Rye can be planted after another cereal crop. Since it is not winter killed, it needs to be destroyed the following spring by tillage or herbicides. Rye is sown at the same rate as other small grains (100-120 kg/ha). Rye may or may not have a positive effect on the subsequent potato yield. It was common practice early in this century in sandy potato soils of New Jersey and Long Island to have rye or a mixture of rye and vetch turned in before potatoes. Rye's role was to maintain tilth in these soils, nitrogen being furnished by mineral fertilisers. Plowing was done early in the spring it was found that a small amount of rye turned under every year was better than a large amount less frequently.

Rye is also a good cover crop after a late potato harvest. Rye has an unmatched ability to grow in the low fall temperatures. It provides a protective soil cover for the winter that is cheaper than hay mulches. In northern regions, rye should ideally be sown in early to mid-September, while in more favourable regions it can be sown up to early October. If sown later, higher seeding rates should be used. An alternative to drilling rye after the potato harvest is to broadcast it before digging the potatoes. If rye is going to be harvested the following year, it is recommended to chop the straw after harvest and incorporate it well. Rye seeds left behind will germinate and make a nice cover crop. Another option is to frost-seed legumes in the rye in the spring.

Use of Spring Cereals

Spring cereals are more often used as rotation crops than as cover crops with potatoes. However, oats, barley or even wheat can be sown as inexpensive catch crops after potato harvest. In contrast to rye, they die in winter, which could be an advantage. If a spring cereal is used as rotation crop sown in spring and harvested in August, seeds that are left behind will often provide a nice cover crop. Barley is not the best choice as it was shown to increase scab levels in the following potato crop.

Use of Grasses

Japanese millet, sorghum/sudan grass or ryegrass can be used as full season cover crops and can also be sources of forage. Most often, these crops are grown to maintain the soil humus content which is often rapidly destroyed in potato culture. The disadvantages of Japanese millet and sorghum/sudan grass are that you need to wait until all danger of frost is over to plant them. While waiting, erosion can take place. Annual soil coverage is therefore not very good with this option.

Crucifers Cover Crops

Cruciferous cover crops are plants of the cabbage family like white mustard, oilseed radish and canola/rape. They are

commonly used as cover crops in potato farming, especially in Eastern European countries. They can be planted soon after cereal harvesting or after early potato harvest. The crucifer is either unfertilised, or fertilised prior to planting with manure or even mineral fertilisers. Crucifers offer several advantages. They grow quickly, resist cold fall temperatures, are excellent nitrogen accumulators, their seeds are cheap and they are winter-killed. They also suppress nematodes. Moreover, mustard has proven to have a suppressive effect against scab and rhizoctonia. One possible disadvantage is phytotoxicity in a cool spring.

Cruciferous cover crops generally increase dry matter content of the following potato crop. In the Ukraine, rape or radish cover crops have been found superior to a peas-oats mixture before potatoes. In Poland, successful experiments have been done where part of the nitrogen normally applied to the following potato crop is applied to rape the fall before. The cover crop then contributes significantly to potato nutrition. In Germany, a system has been developed where the land is prepared for the next potato season after the cereal harvest, and the ridges formed and ready for planting. A cover crop of oilseed radish or mustard is planted along the ridges to provide a soil cover during the winter. In the following spring, the potatoes are planted with the frost-killed crucifers still covering the soil. Potato yields have improved with this system.

Crucifers can be planted after an early potato harvest. They effectively prevent nitrate leaching while controlling weeds and protecting the soil. Being winter killed, they should ideally be tilled into the soil the following spring. In dry conditions, such as on the Prairies, crucifer residues sometimes cause while decomposing phytotoxic effects on the following cereal crop. If crucifers are planted late, heavier seeding rates should be used. If the fall is unusually warm, mowing at flowering stage may be necessary to prevent mustard from going to seed.

Other Cover Crop

Phacelia is another option but is more expensive to sow than the plants discussed previously. It grows like spring cereals but can be sown as a fall catch crop after cereal or early potato harvest. Sunflowers is a sturdy cover crop and is very good against weeds, but it requires heat and is therefore more appropriate in warm regions. Buckwheat is fast growing but can hardly be used as a catch crop because it is easily killed by frost. It is therefore more useful as a rotation crop or as a full-season cover crop. For example, for a good field cleaning, two or three successive sowings of buckwheat in the same season can be done. The risk of self-seeding is high with buckwheat.

Mixtures of Cover Crop Species

Considerable research work has been done on mixtures of various cover crop species. Apart from reducing weed problems, mixtsures are generally superior to single species cover crops on the following potato crop yields. For example:

— A rye/vetch mixture is superior to either rye alone or rape alone;

— A mustard/vetch/peas mixture is superior to mustard alone;

Some mixtures are also better or equal to others. For example:

— A rape/radish mixture is superior to an oats/peas mixture;

— A rye/vetch mixture is equal to a rye/rape mixture;

— Rye/Vetch or rape/vetch mixtures are an excellent choice with a light (10 t/ha) manure application.

A major study is being conducted at the University of Maine using a two-year rotation consisting of potatoes followed by a full season cover crop of oats-pea-vetch (50lbs-100lbs-30lbs/acre). This mixture provides up to 250 lbs. of nitrogen/ac for potatoes.

Interseeding Cover Crops

Systems of potato production with interseeding are being studied at Cornell University. The first results are not too promising. Reduced potato yields were observed with interseeded cereals and/or legumes. Moreover, an adequate weed control could not be achieved with the interseeds only, whether they were sown 3, 4, or 5 weeks after the planting potatoes. Herbicides that would not kill the interseeded plants had to be used to achieve an acceptable level of weed control. It therefore appears that interseeding is not appropriate for potatoes.

Cover Crops to Protect Soil from Erosion after Potato Harvest

A working definition of a cover crop is any crop grown primarily to provide ground cover to improve soil properties, rather than providing a harvestable yield. It is normally grown in the time period between regular crops, although it may be seeded into the preceding crop. Nearly any crop can be used as a cover crop, but some are more useful than others.

The prime advantage of growing a cover crop is to keep the ground covered, as bare soil is an invitation for erosion by wind and water. Another benefit is the cover crop's effect on soil structure and tilth. The improvement in the stability of soil aggregates and water-holding capacity creates a better medium for crop growth. Crops with fibrous roots, such as the grass family, will have a more immediate effect on soil surface structure.

Potato Land After Harvest

- — Drilling or broadcasting rye before Sept. 20 will provide good erosion protection the following spring.
- — A drilled stand of rye is more uniform than the broadcast-incorporated stand and more likely to produce grain.

— Erosion-control benefits of cover crops will increase the earlier the crop is established.

— The establishment of a broadcast-incorporated stand can be improved by increasing the rate of seeding or discing twice to incorporate the seed.

Erosion in Potato Land

Crop Residue Management—The following practices are recommended if a cereal crop is grown before potatoes:

— The combine should have a straw and chaff spreader to distribute residue evenly across the field. Harrowing may be necessary to help spread the straw and chaff.

— Fall tillage of a light texture soil should result in no less than 65 per cent ground cover. One single pass of a heavy-duty cultivator can reduce surface cover by about 20 to 30 per cent.

— Strip Cropping - (alternating strips of potatoes and a crop less prone to erosion) reduces the unsheltered width of erosion susceptible field. Crop strips should be perpendicular to the prevailing winds. The width of the strip will depend on the size of equipment used in cropping and the erodibility of the soil. Recommended strip widths vary from 76 feet in loamy sands to 300 feet in silty clay loams.

— Shelterbelts - Shelterbelts should be planted perpendicular to the prevailing winds. A distance of approximately 660 feet between rows is recommended. The distance between green ash (within row) planting should be six to eight feet apart.

— Mulch - A mulch can provide the same erosion control benefits as a well-established cover crop. Wheat straw (large round bales) spread over the field with a bale shredder can give good protection to the soil after harvest.

On bare, dry soil, wind erosion can remove significant quantities of topsoil. Depending on the size of soil particles, soil movement by wind can occur three ways:

— Surface creep - the largest soil particles roll or slide along the surface. They generally end up in the roadside ditch or along the fencelines.

— Saltation - soil particles bounce up to 1 metre above the surface, dislodging other soil particles each time they hit the surface. These particles also collect along ditches, fencelines and other wind barriers.

— Suspension - the finest soil particles are lifted into air and transported considerable distances.

References

Bugg, R. L., and C. Waddington, "Using Cover Crops to Manage Arthropod Pests of Orchards - a Review", *Agriculture Ecosystems & Environment,* 1994.

Cederbaum, S. B., J. P. Carroll, and R. J. Cooper, "Effects of alternative cotton agriculture on avian and arthropod populations", *Conservation Biology,* 2004.

Creamer, *et al,* "Mechanisms of weed suppression in cover crop-based production systems", HortScience, 1996.

Dabney, S. M., J. A. Delgado, and D. W. Reeves, "Using winter cover crops to improve soil quality and water quality", *Communications in Soil Science and Plant Analysis,* 2001.

Ditsch, D. C., and M. M. Alley, "Nonleguminous Cover Crop Management for Residual N Recovery and Subsequent Crop Yields", *Journal of Fertilizer Issues,* 1991.

Freemark, K. E., and D. A. Kirk, "Birds on organic and conventional farms in Ontario: partitioning effects of habitat and practices on species composition and abundance", *Biological Conservation,* 2001.

Galloway, *et al.,* "Nitrogen-Fixation - Anthropogenic Enhancement-Environmental Response", *Global Biogeochemical Cycles,* 1995.

Giller, K. E., and G. Cadisch, "Future benefits from biological nitrogen fixation: An ecological approach to agriculture", *Plant and Soil* (Historical Archive), 1995.

Grafton-Cardwell, *et al.,* "Leguminous cover crops to enhance population development of Euseius tularensis (Acari : Phytoseiidae) in citrus", *Biological Control,* 1999.

Haramoto, E. R., and E. R. Gallandt, "Brassica cover cropping for weed management: A review", *Renewable Agriculture and Food Systems,* 2004.

6

Barriers in Special Crop Cultivation

The most common species for annual barriers is silage corn, although sunflowers are also popular. Plant the barriers in the spring or early summer to allow the plants to mature before the time comes to harvest the special crop. Barriers may be planted before, during, or after planting the special crop. For corn, plant the belt before June 30.

When properly used, annual barriers will protect the soil from wind erosion after the harvest of a special crop and before the next crop is mature enough to protect the soil itself the following spring.

— In a survey of special crops producers taken in 1992, more than 75 per cent of the respondents indicated that annual barriers are at least partially effective in reducing wind erosion. Nearly half of the respondents rated them good to excellent.

— 1/32 of an inch, the thickness of a five cent piece, equals five tons of soil per acre.

— When visibility is reduced to 1/2 mile in blowing dust, fields may be losing as much as 3/4 of a ton of topsoil per hour.

— Of the water that falls on cropland, an average of 66 per cent evaporates, 25 per cent runs into ponds or streams and only 9 per cent enters the soil.

— In 1991, one-quarter of the province's bean producing

acreage was located in the Rural Municipality of Dufferin.

Use of Annual Barriers

— With late maturing varieties of potatoes. Late potatoes are harvested too late for a cover crop to become established before freeze-up.

— With sugar beets. Sugar beets are also harvested too late in the year to seed a cover crop.

— With early potatoes as an alternative to cover crops if you foresee a problem of finding time to plant cover crops during or after harvest.

— In addition to cover crops. On especially erodible land, annual barriers will provide protection while the cover crop is becoming established, whether it is potato or pulse crop land.

— With pulse crops as an alternative to cover crops.

— When you have reason to believe that soil moisture at harvest time may not be adequate to germinate a cover crop.

— As a method of trapping snow to enhance spring moisture conditions.

Advantages

— Reduces loss of topsoil.

— Snow catch - one foot of snow translates roughly into one inch of water.

— Provides some protection to the crop during the growing season.

— Modification of the field microclimate can enhance crop growth and yield and conserve moisture.

Disadvantages

— If planted as a separate operation, extra time is required in the spring when time is at a premium.

— Requires special equipment or modification to existing equipment.
— Takes land out of production.
— Divides fields into narrow strips.

Most wind erosion damage occurs during the late winter or early spring, therefore, annual barriers are most effective when planted across the direction of the prevailing spring winds. Spring winds are not necessarily from the same direction as the prevailing wind for the remainder of the year.

The closer the annual barriers can be placed, the more protection they provide. The most effective spacing is about 60 feet. However, to minimise the inconvenience of working around the belts, plan the spacing to coincide with the width of your field equipment.

A two-row corn barrier provides the best combination of effectiveness and efficiency. An eight-foot high, two-row belt will reduce wind speed by 35 per cent for a distance of five times its own height, or 40 feet. Wind speed will be reduced significantly for a distance of nine times, or about 70 feet.

Single row corn barriers reduce wind speed only slightly, providing little in the way of erosion protection. However, single barriers will capture snow, enhancing spring moisture conditions.

A four-row barrier does not reduce wind speed significantly more than a two-row barrier and it uses more land. Two rows of corn use about four acres on an eighty-acre field if barriers are spaced 70 feet apart. Four rows use twice that. However, some producers use three or four rows to ensure continued protection if the outside rows of the belt are damaged by spray drift.

On May 20, 1992, a strong south wind filled the air with dust over the Red River Valley. Soil scientists monitored its destructive force on a field in the Winkler area.

The field was corn, seeded into potato residue on dry, sandy soil. Effective soil cover was negligible and the field was smooth and large. Wind speeds ranged from 50 km/hr to 75

km/hr. Wind-blown soil was collected at two sites and the total soil loss per acre was calculated. Soil loss from the field during this single windstorm ranged from 7.25 to 8.23 tons per acre!

Table 1. Soil Loss By Type (kg/metre width)

Movement	Site 1	Site 2
Creep	7.5	7.3
Saltation	529	507
Suspension	115	225
Total (kg/m)	652	740
Soil loss (tons/acre)	7.25	8.23

— *Creep:* The largest soil particles roll or otherwise move along the ground. They generally end up in the roadside ditch or fenceline.

— *Saltation*: Soil particles bounce and fly near the soil surface, dislodging other particles each time they impact the soil surface. These particles also form banks along fences, roads and other wind barriers.

— *Suspension*: The finest soil particles and organic matter are lifted into the air and can be transported considerable distances.

Cover Crops on Special Crops Land

The sole purpose of a cover crop is to protect the bare, erosion-prone land from the wind. They are usually seeded after the harvest of a low-residue crop such as potatoes or in the spring to protect the soil until the new crop has grown enough to protect the land.

Fall rye is the most commonly-used cover crop, although other cereals such as wheat, barley, and oats are also used. A fall cover crop must grow enough before freeze-up to provide protection to the soil.

The latest recommended seeding date for fall cover crops will vary depending on the weather, but used September 15th

as a general guideline. Fall rye is often left for harvest the next summer - as part of the crop rotation. If you are planning to allow fall rye to survive as a crop the following spring, adjust your seeding deadline accordingly.

Cover crops have the added benefit of capturing snow over the winter to enhance moisture conditions in the spring. One foot of snow collected equals approximately one inch of water.

Fall cover crops are best suited for potato land or pulse crops. Sugar beets are harvested too late in the year for a cover crop to become established.

Seed with a Fertiliser Broadcaster Prior to Harvest

This method can work well on potato fields because of the significant amount of soil disturbance associated with the harvest. The cover crop seed is incorporated by the potato harvester and the operation used to anchor the potato vines after harvest. Seeding Rate - 2 bushels/acre because some seed will remain on the surface and some will be buried too deep to germinate.

Advantages

- — uses moisture brought to the surface by the potato harvest
- — relatively fast operation
- — early seeding allows the cover crop more time to establish before freeze-up
- — don't have to take valuable time during harvest to seed

Disadvantages

- — Requires a higher rate of seed application than post-harvest seeding

Seed After Harvest

On potato fields, seed the cover crop within 12 hours of the harvest to take advantage of the moisture brought up by the

harvesting operation. Some operators seed their cover crops at the end of each day of harvesting.

A press drill is preferred because of better packing and fewer problems dealing with potato vines. Hoe drills and air seeders have a tendency to plug up with vines. On pulse crop fields, cover crops may be seeded with any one of a variety of seeders but packing is important to obtain a good seed-to-soil contact in soil that is likely to be dry. In-line packers are the most efficient.

Seeding Rate - 3/4 to 1 bushel/acre

Advantages

— drilling the seed into the soil rather than broadcasting it results in better germination and allow lower seeding rates

— on potato land, it uses moisture brought up by harvesting

Disadvantages

— dealing with potato vines

— takes more time and horsepower than broadcasting seed

An Aircraft prior to Harvest

Seeding cover crops with an aircraft can be effective and economical, especially where large areas of land are involved. Costs per acre for seeding from the air are comparable to other methods when considering the operator's time, operating costs, and wear and tear on machinery.

Advantages

— Time. The producer doesn't use valuable time seeding and a lot of acreage can be seeded in a short time.

Disadvantages

— Uneven distribution of seed among hills and furrows may be a problem.

Anchoring Potato Vines

If potato vines are not anchored, they will roll up and blow away. Use a disc with two-thirds of the discs removed to anchor vines. Have it set up for burying the vines, not for tillage.

Travel slowly - about 3 miles/hr - in same direction as the rows or diagonal to them. The operation will leave some cover and ridge the soil. Using a vine chopper attachment on the potato harvester eliminates the need to anchor the vines.

Removing a Fall-planted Cover Crop

Normally, a cover crop will be killed by frost before it grows too much to be a problem. However, if the cover crop's use of soil moisture and the potential amount of straw on the field the following spring are becoming concerns, use a herbicide to kill it. Fall rye may be left until spring, then burned off with a herbicide either before seeding a new crop or pre-emergence.

Planting the New Crop

Direct seed the new crop into the cover crop residue. In the case of fall rye, burn off the new growth before seeding or immediately after seeding.

References

Brook, R. C., R. J. Fick, and T. D. Forbush, Potato Storage D*esign and Managemen, East Lansing, MI.: Michigan State University* Agricultural Engineering Information Series, AEIS#627, 1995.

Christ, B. J., Potato Diseases in Pennsylvan*ia,* Pennsylvania College of Agricultural Sciences, University Park, PA. 1998.

Celis, C. *et al.*, "Environmental biosafety and transgenic potato in a center of diversity for this crop", *Nature,* 2004.

Jackson, S.A. Hanneman Jr., R.E., "An assessment of crossability between cultivated potato and its wild relatives", *American Potato Journal (USA),* 1996.

Frederick, R.J. Virgin, I. Lindarte, E. (eds.), *Environmental concerns with transgenic plants in center of diversity: Potato as a model, proceedings,* Stockholm (Sweden), Stockholm Environmental Institute (SEI), Biotechnology Advisory Commission (BAC) Inter-American Institute for Cooperation on Agriculture (IICA), 1995.

Sieczka, J. B. and R. E. Thornton, *Commercial PotatoProduction in North America*, Orono, ME: The PotatoAssociation of America Handbook, 1992.

Weisz, R., Z. Smilowitz, M. Saunders, and B. Christ, *Integrated Pest Management for Potatoes*, University Park,PA: Penn State Cooperative Extension, 1995.

7

Seed Potato Production

The range of potato varieties produced in the Prairie Provinces varies with the demand and ultimate use of the raw potato. Most of the fields entered for certification are for the production of processing potatoes. Russet Burbank and Shepody are the main French fry processing varieties, while Atlantic, Norvalley and Snowden are the main chip processing varieties.

The majority of these seed potatoes are grown under contract with potato processing companies. Other varieties are produced for the production of table potatoes. These include red skin, white skin and yellow flesh varieties.

Seed Potato Act

In Canada, the inspection and certification of seed potatoes is governed by legislation contained in Seeds Act and Part II of the Seeds Regulations. The seed potato regulations define seed potatoes as: "potato tubers certified pursuant to Part II of the Seeds Regulations of the Seeds Act for reproduction purposes". Seed potatoes must, therefore, meet the standards set out by these regulations. Potato tubers that do not meet the seed potato standards do not qualify as seed potatoes and are defined as non-certified potatoes or table stock potatoes. Seed potatoes, which have been transported without official tags or a bulk-movement certificate, are also defined as non-certified potatoes.

The Canadian Food Inspection Agency administers the Seeds Act and Part II of the Seeds Regulations and ensures that a uniform standard is applied across the country. CFIA headquarters in Nepean, Ontario, assumes the main administrative role for developing new policies and regulatory amendments in consultation with a national seed potato industry stakeholders committee. The Operations directorate of CFIA is responsible for seed inspections and issuing certification numbers and phytosanitary certificates.

CFIA Officers and Inspectors

- To approve or reject grower applications for crop inspection
- To inspect and verify the cleaning and disinfecting of equipment and storage used in seed potato production
- To inspect grower crops and to record and report the inspection results
- To pass, downgrade or reject the certification of fields entered for inspection
- To issue Growing Crop Certificates to fields passing inspection
- To issue Rejection Notices to fields failing to pass field inspection
- To revoke certification on crops which fail to continue to meet the certification requirements
- To inspect and verify that seed potatoes have been graded prior to shipment
- To issue official tags and bulk-movement certificates on seed being shipped
- To carry out tuber inspections when a request for re-inspection of a lot has been received
- To perform post-harvest inspections of winter grow out plots under disease evaluation
- To accredit qualifying laboratory testing for seed potatoes

Responsibilities of the seed potato growers

- — To apply for inspection on the official form
- — To ensure that all equipment and storage facilities are cleaned and disinfected before being used in the production and storage of seed potatoes
- — To ensure that only eligible seed potatoes are planted on the seed farm
- — To ensure the crop is protected from varietal mixture and disease contamination
- — To ensure the growing crop is well managed
- — To rogue diseased and off-type plants from seed fields
- — To ensure the harvested crop is stored in a manner that prevents varietal mixture, contamination by disease, loss of identity, and tuber deterioration
- — To grade the seed potato tubers to regulatory quality standards and the prescribed size requirements prior to shipping
- — To correctly label all seed potatoes sold using official tags or bulk-movement certificates
- — To keep field rotation, varietal history, production & storage figures and sales records to document product activity

Seed Potato Production

Seed potato production in Canada uses a limited generation system, which means that seed passing inspection must advance to a lower class with each generation of production. A seed crop for each production generation cannot be maintained at a specific class or move to a higher class. Nuclear seed is at the top of the seed potato production pyramid is Nuclear Seed. Nuclear seed must be produced from disease free parent stock in a protected environment.

The parent stock has been extensively tested to confirm the absence of potato pathogens. Nuclear stock production is subjected to a series of laboratory tests to confirm freedom

from certain potato pathogens. These pathogens include: potato virus X (PVX), potato virus S (PVS), potato virus M (PVM), potato virus A (PVA), potato mosaic virus (PVY), potato leaf roll virus (PLRV), potato spindle tuber viroid (PSTV), and bacterial ring rot (BRR).

In the field, seed can be increased for seven years. There are therefore seven sequential classes of seed potatoes in the Canadian Seed Potato Certification system. Each class has a defined tolerance level for the presence of certain diseases and foreign varieties. All classes must be completely free of spindle tuber and bacterial ring rot. The tolerance levels for the other diseases increase slightly with each class produced. Therefore, the class designation describes a precise seed quality based on parentage and the amount of disease detected in the growing crop.

Parallel to the schedule of classes is a field generation designation. Contrary to the class designation, the field generation does not involve disease tolerance levels and is not a quality designator. Generation is a term that identifies only the number of years that the seed has been grown in the field or the number of field production generations after Nuclear Seed.

New Seed Grower

Anyone interested in becoming a seed potato grower should contact the local office of the CFIA. This should be done at least three months prior to receiving seed supplies for the spring planting. An application made by a new grower will only be approved when cleanup and an inspector of CFIA has verified disinfection of the premises and equipment. This must be carried out before the grower can receive seed potatoes. A new applicant should also be aware that land used for seed potato production must not have been in potato production during the previous two years.

Crop Inspection

Applications for field inspection should be made on forms supplied by CFIA. The completed application, payment of

assessed fees and other required information should be delivered to the address given in the application package for all interested seed potato growers. The application deadline for seed potatoes is June 15 of the current crop year. The following information must be provided for each potato field planted:

- — Variety
- — Hectares
- — Seed Potato Class and certification number
- — Weight planted (in cwt or kg)
- — Location of the field
- — Previous crop grown on the field

The CFIA must be made aware of all crops planted by a seed potato grower. All potato fields planted by a seed potato grower must be planted with eligible seed. This includes Foundation or a higher-class seed, or if U.S. seed is planted it must be of a class equivalent to Foundation or higher.

Certified class or non-certified potatoes cannot be planted on a seed potato farm. US seed potatoes must be tested for Bacterial Ring Rot in a Canadian accredited laboratory to qualify for official Canadian certification status. Additional documentation is required when the applicant has planted seed potatoes originating from another farm. This documentation is required to provide evidence of seed eligibility. It also provides evidence of seed origin and of the quantity of seed planted. The necessary documentation includes:

- — An official tag from one container of those potatoes purchased in containers
- — A copy of the bulk-movement certificate in the case of seed potatoes purchased in bulk
- — A copy of a nuclear stock certificate in cases where nuclear stock is planted
- — An Import Permit from CFIA headquarters and a Phytosanitary certificate from the USA are necessary to import seed potatoes. In addition, for certification

eligibility a Bacterial Ring Rot test from an accredited Canadian laboratory is required and a North American Seed Potato Health certificate from the state of origin is recommended to determine class eligibility of the seed potatoes in the Canadian system

A map of the exact location of each field should also be included with the application. The map should include as much information as possible about each field. Growers should also identify the access route to each field.

Conditions of inspection

The following conditions may result in the rejection of a grower's application:

- — The total area entered for certification is less than 0.25 acres (or 0.1 ha) in size
- — Bacterial ring rot infected potatoes were found in the field, buildings, or on equipment used by the grower
- — An inspector did not verify cleanup and disinfection of the buildings and equipment
- — Non-eligible potatoes have been planted on the farm. The following are identified to be non- eligible potatoes:
  - — Certified class seed potatoes
  - — US seed potatoes that do not have any Canadian class equivalency
  - — Non-certified (table) potatoes
  - — Seed potatoes exposed to contamination by ring rot
- — The crop is growing in a field where there has been an occurrence of bacterial ring rot. (Unless the inspector has verified that the field has been free of potatoes, including volunteers for the previous two years)
- — The certification fee has not been paid

An inspector may refuse to inspect a specific crop or field where:

- — A condition exists that interferes with an inspector's ability to conduct a visual inspection of the crop. This

can include poor growth, lack of vigour or leaf injury brought about by late planting, lack of cultivation, the existence of excessive weeds, weather conditions, soil conditions, chemical injury, insect damage or pesticide damage

— The field is located within 197 feet (60 metres) of another field that has visible symptoms of a virus disease that exceeds the tolerance limits for the class of seed being produced

— The distance between adjacent seed potato fields is less than one blank row

— The crop is growing in a field where non-certified potatoes were planted in the previous two years

The seed producer may be requested to submit all the official tags from the containers of seed potatoes that were purchased. Potato fields entered for certification are inspected two or three times during the growing season by a CFIA inspector, prior to top-kill. At the time of inspection, the inspector will decide whether the crop meets the standards for the class to be produced. Before the inspection occurs, the seed grower should ensure that the seed fields are rogued of diseased plants. Potato fields that are not entered for inspection but are being grown by the seed grower for non-seed use may also be subject to a field inspection.

Certified Seed Potatoes

The following are the final inspection tolerance levels for Elite 1, Elite 2, Elite 3, Elite 4, Foundation and Certified class seed: Nuclear and Pre-Elite class seed have a 0.0% tolerance for all diseases on final inspection. Finally, a random tuber sample of Elite 2, Elite 3, Elite 4 and Foundation seed lots are subjected to a laboratory test and must be found Not Popitive for the presence of Clavibacter michiganensis subsp sepedonicus (Bacterial Ring Rot) before it can be sold. Where an inspector has inspected crops in a seed potato farm and finds they meet the standards set out in the regulations, a "Growing Crop

Certificate" will be issued to the grower. The Growing Crop Certificate specifies the following for each crop:

— Class and variety
— Number of hectares that passed inspection
— Certificate number

Crop certificates will not be issued or may be revoked in the following cases:

— An inspector learns that any one of the lots are infected with Clavibacter michiganensis subsp sepedonicus
— An inspector has determined that the crop does not meet the standards for any class of seed
— An inspector has determined that the crop has been infected with the potato spindle tuber viroid
— The crop has been treated with or exposed to a sprout inhibitor
— Any one of the lots has come into contact with used, shared or custom equipment that was used in the harvesting, storing or grading of a lot that was infected with ring rot bacteria
— The seed potatoes have lost their identity
— The seed has lost its ability to propagate
— The inspector determines that non-certified potatoes from another farm are stored with the certified seed potatoes or that the certified seed potatoes were stored, graded or handled with common equipment

Seed Potato Production Practices

Many of the practices used in seed production are similar to those used in commercial production.

Seed Selection and Handling

A potato crop is only as good as the seed. When purchasing seed, ensure the supplier has a reputation of delivering a quality product. Also, check that the growing area has a history of producing quality seed. A visit to prospective seed

suppliers both during the growing season and once the crop is in storage will assist to ensure quality. Before buying any seed for re-certification, the grower should be satisfied that the seed is going to have a good chance of passing the certification inspections. It is generally fruitless to plant a seed lot having even a low incidence of disease with the expectation that the diseased plants can be removed through roguing.

Check the results of field and storage inspections and insist on Post Harvest test results before agreeing to purchase seed. When negotiating purchase of a seed lot, determine the method of delivery, date of delivery, desired tuber size distribution, tuber temperature and degree of sprout development. Inspect the seed immediately during delivery to ensure that it meets the agreed upon quality stipulations. The buyer should contact both the seed potato grower and CFIA within two working days after receiving the seed potatoes, if not satisfied with the quality.

If a re-inspection is requested it will be carried out by a CFIA inspector within five working days after receipt of the complaint. The buyer should have seed tags and some unopened sacksor the bulk-movement certificate on hand, so that the proper information can be conveyed to the inspector when lodging a complaint. All machinery, transport and storage surfaces that the seed will contact should be cleaned and disinfected prior to receiving new seed.

Sanitation consists of cleaning and disinfecting all equipment, storage, tools and pallet boxes that contact the seed potatoes. Since most disinfectants are inactivated by soil and plant debris, it is essential that this material be removed by thoroughly cleaning the equipment and storage with a pressure washer or steam cleaner before the disinfectant is applied. The seed storage and all equipment coming in contact with the seed lot should be disinfected with a quaternary ammonium compound such as Ag-Services Incorporated General Storage Disinfectant, Bardak 2210 Disinfectant Sanitizer, or DMR-23 Disinfectant.

Surfaces must remain wet for at least 10 minutes for the disinfectant to destroy disease organisms. Restrict contact

between incoming seed lots and any potatoes left over from the previous year. Cull potatoes are a particularly important potential source of disease. Avoid sharing equipment between farms; sharing seed cutting equipment is particularly risky. Whole seed has many advantages over cut seed, particularly in the production of seed potatoes:

— The wounds produced by cutting seed presents entry points for many disease-causing organisms

— Seed cutting equipment contaminated with a disease pathogen spreads disease both within and between seed lots

— Where the seed regulations require tuber unit planting the planting of whole seed is an acceptable alternative

— Whole seed tends to produce large numbers of relatively small tubers, which is ideal for seed prod-uction

— Whole seed generally produces a more uniform stand both in emergence percentage and in growth characteristics, which makes more efficient use of the production inputs

— Whole seed eliminates the equipment and labour cost of cutting

Timely planting is as important in seed production as in table or processing production. Plant early and plan on an early vine kill and harvest. This will reduce the risk of introduction and spread of viral diseases by late season winged aphid. Plant the highest generation seed lots first to minimise risk of disease. Clean all equipment between seed lots. Cup type planters are less likely to spread disease within a seed lot than pick-type planters.

In-row spacing is the main factor controlling tuber size. Optimum seed spacing is based on varietal characteristics and climactic conditions. Closer in-row spacing is used in seed production to maximise yield of the desirable, smaller tuber sizes. In-row spacing of 6-12" (15-30 cm) are recommended for seed production, depending on variety, irrigated or rain-fed production, area in which the potatoes are grown and other climactic conditions

Fertility requirements for seed are generally comparable to other types of production. Highly vigorous, early-generation seed may require less fertiliser for optimum yields, particularly if the crop is grown under rain-fed conditions or is going to be top-killed early.

Weed control Uncontrolled weed growth in seed potato fields will:

— Interfere with proper inspection and roguing
— Restrict air flow through the canopy which increases the potential for disease development
— Provide alternate hosts for diseases and insects capable of spreading disease
— Rob the crop of fertility and moisture
— Interfere with a timely and gentle harvest

Effective weed control in seed fields can be obtained through timely tillage and the application of approved herbicides. Note that some herbicides registered for use in table and processing potatoes are not recommended in seed crops. Caution should be exercised when applying herbicides to adjacent crops. Drift of certain herbicides can be translocated to the daughter tubers. The affected tubers lack vigour and produce plants displaying symptoms of herbicide injury. Herbicide residues from previous crops can also be taken up by the seed crop producing similar results.

Disease Prevention, Roguing and Insect

In seed potato production it is imperative to maintain disease within levels tolerated by the certification system. Failure to achieve the tolerated disease levels will result in either the downgrading or rejection of specific seed lots or in certain cases the rejection of the entire farm from certification. Bacterial ring rot is the most devastating disease since its presence will result in the rejection of the total farm. Many disease pathogens are normally present in potato production areas.

Disease and varietal mixture	E1	E2	E3	E4	FD	Cert
PSTV	0.0	0.0	0.0	0.0	0.0	0.0
Ring rot	0.0	0.0	0.0	0.0	0.0	0.0
Total all viruses	0.0	0.1	0.2	0.3	0.5	2.0
Total Blackleg and wilts	0.1	0.2	0.3	0.5	1.0	2.0
Varietal mixtures	0.0	0.0	0.0	0.1	0.2	0.5

Consequently, there is a constant potential for the introduction of these pathogens into a seed crop. The seed grower must be constantly aware of the potential sources of disease introduction, the mechanisms of disease spread and the methods of preventing disease introduction. Grower reputation is built upon preventing or minimising disease introduction and spread in the seed crop. Preventing disease contamination of seed lots is the main objective of seed potato production. Disease contamination can occur from four sources:

— The parent seed used to produce the crop
— The equipment and storage used in potato production and handling
— The soil used to produce the crop
— Insects transmitting disease from other potatoes in the crop production area

Many disease problems in a seed operation can be traced indirectly and sometimes directly to purchased seed. This is especially true with bacterial ring rot, although leaf roll and mosaic viruses are also frequently problems in purchased seed. Purchase seed from experienced growers who have a reputation for producing quality, disease free seed. Contaminated potato handling equipment is an important source of disease introduction and spread. Ring rot bacteria can contaminate equipment used in fieldwork or potato handling and potato structures through contact with diseased material. Dormant ring rot bacteria can survive several years in a desiccated state on the surfaces of equipment or storage structures. For these reasons, seed growers are discouraged

from purchasing used equipment or from sharing equipment with other producers.

If this type of equipment is used it must be thoroughly cleaned of dirt and debris with a high-pressure washer then disinfected under the supervision of a seed inspector. Parts with the potential to harbour disease inoculum such as the sponge or soft rubber rollers used on seed cutters, should be replaced. When equipment is in need of a cleanup, the seed potato growers must use proper sanitation procedures and apply them rigorously. Sanitation consists of cleaning and disinfecting all equipment, storage, tools and pallet boxes that contact the seed potatoes.

Since most disinfectants are inactivated by soil and plant debris, it is essential that this material be removed by thoroughly cleaning the equipment with a pressure washer or steam cleaner before the disinfectant is applied. The seed storage and all equipment coming in contact with theseed lot should be disinfected with a quaternary ammonium compound such as Ag-Services Incorporated General Storage Disinfectant, Bardak 2210 Disinfectant Sanitiser, or DMR-23 Disinfectant. Surfaces must remain wet for at least 10 minutes for the disinfectant to destroy disease organisms.

Certain potato diseases can survive from season to season in the field. Depending on the type of pathogen, it may survive in the resting form either in the soil or in potato plant debris, or in a living form in surviving potato tubers. On occasion, diseased tubers survive the winter and grow the following spring as diseased volunteer plants. These volunteer potatoes are a source of contamination for the current season crops. Most seed potato producers practice a three to four year rotation to minimise soil disease problems. Fields with a previous history of bacterial ring rot infection should not be used for potato production for a minimum of two years, and during those two years volunteer potato plants must be eliminated.

Sucking and chewing insects carry many viral diseases of potatoes. For example, aphids transmit potato leaf roll and mosaic viruses from plant to plant. Aphids become infected

with the virus while feeding on diseased plants and are capable of transmitting the virus to healthy seed plants. Attempts to control aphids in the seed fields may fail to prevent disease introduction. A virus-infected aphid may feed on a potato plant leaf and introduce the virus into the plant before it succumbs to an insecticide; however, insecticides do prevent disease transmission from continuing unchecked.

Seed growers who are producing in commercial production areas with the associated greater risk of disease inoculum should attempt to isolate their seed fields from potential sources of contamination. If that option is not available, then the grower should attempt to control viruliforous (virus-infected) aphids in the source field before winged aphids migrate to seed fields, spreading virus to the seed crop. If the seed field already has virus-infected plants then the application of pesticides will help prevent spread within the field. Planting seed with low or zero incidence of virus as reported by a post-harvest test, and limiting the number of seed generations produced in areas of commercial production will help to limit virus spread within seed crops. Roguing all infected plants as early as possible prevents in field spread by aphids.

Purpose of Roguing

The purpose of roguing is to:

- Remove plants of a variety different from that planted in the field
- Remove diseased plants that will produce diseased seed tubers or which represent inoculum sources for disease spread within the field
- Ensure the seed field passes inspection.

Although roguing is a common practice in seed potato production it is only practical when the field is contaminated with a very low incidence of an off variety or diseased plants that are easily distinguishable. When rouging, remove plants, seed piece and daughter tubers from the field. The process of roguing requires:

- — Experienced individuals, who are capable of recognising the symptoms of the important potato diseases
- — Appropriate timing
- — Environmental conditions which contribute to the expression of visible disease symptoms

In addition, roguing is costly, as the field may have to be walked several times to remove the off variety or diseased plants. Roguing for viral diseases is best done prior to flowering and during overcast weather. Although roguing is considered to be a valuable tool in the maintenance of seed quality, it is rarely completely effective. Only those plants that are manifesting visually recognisable symptoms can be rogued. Symptom expression can be influenced by many factors including environmental conditions and varietal characteristics.

In addition, not all pathogen-infected plants express disease symptoms to the same degree or at the same stage of growth. The end result is that much of the roguing is conducted to provide a visually clean field that will satisfy the seed certification specifications. Tuber-unit planting is a tool that has been used to improve the effectiveness of the roguing process. All seed pieces cut from a single tuber are manually fed into a planting mechanism, which places them consecutively in a row.

A gap is sometimes left between the seed pieces generated from different tubers. If the plants originating from any of the seed pieces cut from a single tuber express symptoms of disease, then all the seed pieces cut from the tuber are removed from the field. This ensures that all seed is removed from the field even if only one piece is manifesting symptoms of disease or has characteristics of a foreign variety. Since tuber-unit planting is slow and very labour intensive, its use is limited to small plots of high value seed. This practice was a regulatory requirement for the first and second field generations of seed. Regulatory amendments approved in May 2002, resulted in the removal of this requirement from

seed regulations, due to cost, and time consideration, relative to industry benefits. However, it is considered an important management tool for seed growers, particularly when a grower has concerns for the level of virus that may be present in seed planted.

Certain types of aphids and leafhoppers have the potential to both introduce and spread viral diseases within a seed crop. Controlling the populations of these insects by the application of insecticides at planting or after crop emergence will reduce, but not completely eliminate, problems with insect-borne disease. Seed growers should monitor insect populations and if possible, plan on vine killing before insect numbers increase beyond acceptable levels. Isolation and planting clean seed will also reduce the risk of insects spreading disease within the seed crop.

A continuous supply of an adequate amount of moisture via timely rainfall or irrigation is crucial for maximising yields and quality of seed potatoes. Adequate moisture at tuber set is particularly important as it encourages setting of a large number of tubers and controls or reduces the severity of common scab. Over-irrigation increases susceptibility to late blight, blackleg, powdery scab and tuber rots. The wheels on wheel-move and centre pivot irrigation systems may contribute to the transmission of some viral or bacterial diseases within the field.

Harvest and Storage Management

Vine Killing: The purpose of vine killing in seed production is to:

- Reduce the risk of late season virus transmission by winged aphids. Early generation seed should be killed by mid August
- Stop tuber growth to obtain a desirable tuber size profile
- Promote skin set and prevent bruising. A well-developed skin resists damage during harvest, which reduces the potential for infection to enter the tuber

causing rot. Vines should be killed for 10 to 14 days before harvesting to achieve maximum resistance to bruising and skinning

— Allow easy separation of the vine from the tubers during harvest

Growers must ensure that each class and variety of seed is kept separate through harvest, grading and storage. Bin dividers can be used to subdivide larger bins, allowing the safe storage of multiple seed lots. The contents of each bin should be clearly labelled to avoid accidental contamination with differing varieties or generations. Seed cannot be treated with storage-applied sprout inhibitors such as CIPC. Due to the potential for drift, seed potatoes cannot be stored in the same building as any potatoes treated with CIPC.

Similarly, seed should not be stored in buildings treated with CIPC in previous years as this volatile compound is absorbed into the storage structure. Seed should be cured at 55-60° F (13-15° C) and 95% RH with good airflow for two weeks. Once curing is completed, reduce the ventilation rate and gradually cool the pile to a final holding temperature of 37-40° F (3-5° C) with 95% RH. The condition of the pile should be checked regularly during the storage period. Restrict access to the storage to reduce the potential for introducing disease.

Post Harvest Testing

The purpose of the post harvest test (PHT) is to preview a seed lot's disease status for the next crop year. Late season virus transmission by aphids is difficult to detect in field inspections as they seldom produce recognisable symptoms. A reliable PHT identifies if virus infections have occurred and enables the grower to determine the crop's suitability for seed purposes. The PHT is performed on a random sample of seed tubers collected by the seed grower.

The PHT is not part of the Canadian Seed Potato Certification programme. It is an industry-administered programme designed to provide a level of assurance of seed potato quality, which is not achievable through the official certification programme. Participation by growers in the PHT

is voluntary. The value of the PHT is underestimated, since the test is not considered a condition of certification. Nevertheless, it is suggested that all seed lots and fields should receive a PHT. There are three options for PHT available to the industry. These include:

— Visual inspections of the growing plants at southern grow-out sites (California, Florida or Hawaii)
— Visual inspections at the southern grow-out site combined with a laboratory test (E.L.I.S.A.) on leaf material collected at the southern site
— A laboratory test (E.L.I.S.A.) of leaf tissue collected from plants produced in a growth chamber or greenhouse from tuber samples received from the grower

The reliability of the PHT results depends upon the following:

— The quality of the sample collected.
— The post harvest test method. Both the grow-out and laboratory PHTs are capable of detecting viruses, but only the grow-out test can detect foreign varieties. The grow-out test is based on a visual assessment of disease symptoms. Expression of visual symptoms is subject to many external factors, which limits reliability. The laboratory test (E.L.I.S.A.) is more reliable at quantifying virus levels because it is not affected by external factors. Certain varieties such as Shepody and Russet Norkotah normally mask the field symptoms of mosaic. A laboratory PHT for PVY should be a standard practice for these cultivators

Seed potato tubers for Post Harvest Testing (PHT) and for BRR testing must be selected at random to provide an unbiased sample that is representative of the whole field or lot. To ensure the samples submitted for testing truly represent the seed crop, it is suggested that seed growers collect a field sample of about 1000 tubers from each field. A test sample can then be randomly collected from the field sample.

The most efficient method of collecting a representative tuber sample from the crop is to collect the field sample during the harvest operation when all tubers are equally accessible.

To ensure a representative sample is collected, one farm staff member should be given the training and responsibility for collecting the tubers. The sample collector should be instr-ucted to collect a uniform number of suitably sized tubers (1 to 2 ounces (30-60 g) from every load delivered to the storage. The number of tubers collected per truck arriving at the storage will depend on the size of the field being harvested.

When the harvest of the field is complete, the field-samples should be bagged, labelled, and placed into proper storage for safekeeping. This field sample will be later sub-sampled for the individual tests. Four hundred (400) suitably sized tubers are required for most post harvest tests. These samples are randomly collected from the field-samples.

It is suggested that the remaining portion of the field-sample be held in the storage as long as possible. This material may be useful if for testing for the presence of genetically modified organism (GMO) or BRR. Four hundred (400) 1-2 oz (30-60 g) tubers are required for the grow-out PHT. The total weight of the sample should not exceed 50 lbs (22.7 kg). Note: Samples containing tubers in excess of 3" (7.6 cm) in diameter may be rejected for the grow-out PHT. All samples should be packed in new bags and labelled twice with a tear proof label providing the following information:

— The grower's name
— The Variety
— The Class
— The Certification Number for current production year
— Grower field numbers

Tubers submitted for the laboratory PHT or BRR test could be any size since only portions of the tubers are sent to the laboratory.

The current regulatory requirement for testing for Clavibacter michiganensis subsp. sepedonicus, the causal bacteria of Ring Rot, is that all certified seed potato lots except for Pre-Elite, Elite I, and Certified classes, that are sold by a farm unit, must be tested. Contact the Canadian Food

Inspection Agency (CFIA) for a list of BRR accredited laboratories. The size of the BRR test sample required is based on the number of hectares that were entered for certification. Make every effort to prevent bruising. Seed potatoes should be sized to meet the market requirements. The standard method of sizing seed potatoes is based on weight. Tubers can also be graded by dimension. When sizing by dimension, screen sizes appropriate for each variety must be used. All seed potato classes are subject to the same tuber standards.

The quality standards provide a tolerance for symptoms of disease, damaged tubers and varietal mixtures that could affect the quality of the seed. Tolerance for disease and defect in the graded stock is defined by the tuber standards in the seed potato regulation. In addition, the tolerance for foreign varieties is as follows:

— Elite 3 or Elite 4 classes of seed - none

— Foundation class of seed - 0.05%

— Certified class of seed - 0.10%

In any lot of graded seed potatoes, at least 98% of the tubers shall be firm and well shaped. Growers are encouraged to take samples of the graded stock regularly during the grading operation to ensure the grade requirements are being met. Where problems are identified, appropriate adjustments should be made at the grading line. These may include adding additional staff or reducing the speed of the conveyers. Seed potatoes may be packed or shipped in a variety of container sizes, including the normal jute bag and bulk containers. Bagged seed potatoes must be labelled with an official tag that identifies the variety, class and certification number of the seed lot being shipped.

References

Brook, R. C., R. J. Fick, and T. D. Forbush, *Potato Storage Design and Managemen*, East Lansing, MI.: Michigan State University Agricultural Engineering Information Series, AEIS#627, 1995.

Christ, B. J., *Potato Diseases in Pennsylvania*, Pennsylvania College of Agricultural Sciences, University Park, PA. 1998.

8

Pesticide Resistance Management

Selective breeding is an important tool in agriculture. Through selection of important traits, breeders have been able to produce the diversity of high yielding, high quality crops used in food production. Unfortunately, another form of selective breeding is detrimental. Through repeated application of pesticides, producers are inadvertently selecting resistant "breeds" of insects, weeds and diseases. All natural populations have a wide variety of genetic traits. Within a population, there may be a few individuals that are naturally resistant to a particular pesticide. Pesticide resistance develops when a pest population is constantly being selected through repeated use of pesticides with a similar mode of action.

The few surviving individuals multiply and eventually replace the susceptible population. The rate of increase of the resistant individuals within the population depends on the frequency of application and the reproductive potential of the pest in question. Once resistance is established, pesticides within that chemical group are no longer effective. Weed resistance to herbicides is becoming more prevalent on the prairies and may affect weed management in potatoes. Populations of wild oats and green foxtail are known to be resistant to Group 1 herbicides. The potato beetle is notorious for its ability to have resistant populations selected, sometimes within two years after the introduction of a chemical with a new mode of action.

Populations of the potato beetle are resistant to many chemicals, forcing producers to rely on new chemical groups. Within the last several years, populations of the beetle resistant to most conventional insecticides. The first signs of resistance are incomplete control of the pest after pesticide application.

With weeds, this often appears as a healthy weed patch within a field where most weeds have been well controlled. However, before assuming resistance, other factors that could be responsible for pesticide failure must first be ruled out. Such factors include improper application (clogged nozzles, drift etc.), unfavourable weather, and emergence of new pests. If you suspect resistance, then you are strongly encouraged to contact your provincial potato extension officer.

Prevent Resistance to Pesticides

Prevention is the key strategy in resistance management. Once resistance has established, there is no "cure". If available, alternative pesticides must be used. The development of pesticide resistance can be delayed by rotating between different groups (classes or families) of the pesticide, as each group tends to have a different mode of action.

Pests or their offspring that survive an application would be killed by a pesticide from a different group in the following application. The rotation between pesticide groups could occur within the same or following season Pesticide resistance management through rotation of pesticide groups is only possible in situations where totally resistant populations are not yet present. Adequate record keeping of pesticide use is critical when rotating pesticides over years to minimise resistance development. Continued dependence on chemicals as the sole method of pest control is a sure recipe for the development of pesticide resistance.

Agronomic practices can play an important role in managing pests and the development of pesticide resistance. For instance, a cereal crop planted after potato harvest to protect soil from erosion may also aid in controlling weeds.

Crop residues shade weeds that require light to germinate and physically impede growth of young seedlings.

A significant number of beetles will over-winter in the previous field. The further the distance between the current potato crop and the source of over-wintering adults, the less chance the beetles will survive the journey to the new field. Refer to the sections on weed, insect and disease control below for more alternative pest control strategies that reduce dependence on chemical applications. By implementing the following pesticide resistance management practices today, growers will ensure that there are more pesticides to choose from in the future.

— Monitor your fields and apply pesticides only when necessary. Pest populations below an economic threshold can be beneficial, as they provide food for natural enemy populations and also serve to ensure genetic diversity within the population

— Employ agronomic practices that play a role in managing pest populations.

— Apply pesticides only to areas of the field where the pest is a problem

— Alternate between different groups of pesticides. Remember to take into account the pesticides you may wish to apply to control other pests later in the season. For example, if you expect to use an insecticide to control aphids later in the season, then choose an insecticide from a different family to control potato beetles early in the season. Alternate pesticide groups within and between years. Refer to your provincial crop protection manual for a listing of the different chemical groups

— Never repeat a treatment because adequate control was not obtained. This is the first sign that resistance is high. Always change chemical groups if repeated treatments are necessary

— Contact your provincial potato crop specialist if you

suspect resistance is present. The sooner resistance is detected; the earlier alternatives can be implemented

Organic and Pesticide Free Production

Organic potato production prohibits the use of synthetic chemicals, fertilisers, pesticides, growth regulators, or genetically modified varieties. The decision to grow organic potatoes may be based on a personal desire to adopt an organic production system, to capitalise on the demand in the organic marketplace, or for various other reasons.

A wide range of insect, disease and weed pests can attack potatoes. The key to successful production of potatoes without the use or synthetic pest control products is prevention and plant nutritional health. Experienced potato producers will be aware of the major threats to healthy potato production and know the consequences of aggressive pest attacks.

Novice commercial potato producers may not be aware of the destruction of uncontrolled pest invasions and poor plant nutritional health and must be aware of the pitfalls of failing to adequately protect potato crops. Certain pesticides are approved for use by certification bodies. Consult the "Permitted Materials List" of the certifying body before applying a product. Ensure that all products used as pesticides are also approved by the Pest Management Regulatory Agency (PMRA). PMRA is the federal agency that approves the use of all pesticides in Canada. It is against the law to apply a pesticide that is not registered by PMRA.

Overall plant-health considerations

— It is recommended that organic potatoes be grown in a minimum of a 4- year rotation to minimise yield losses from soil-borne diseases such as Verticillium wilt, Fusarium and Rhizoctonia

— General soil fertility must be maintained by a well-planned management system involving rotations, legumes, straw and composted manure.

— Manure plays an important role in providing plant nutrients, however, organic producers must ensure that

the manure has been composted properly and meets the standards approved by the certification authority

— Organic producers cannot use a fungicide seed piece treatment, so it is recommended that whole seed be planted. Whole un-cut seed tubers are less likely to become infected with soil borne diseases than cut seed pieces.

— Plant low generation certified seed, which has a lower incidence of seed borne diseases

— Certain certifying agencies and end-users may require the use of organic seed. Organic producers must clarify this prior purchasing seed

— Vigorously growing potato plants are more resistant to insects and diseases than plants under stress. For example, the main prevention for early blight is ensuring that a potato crop has an adequate, season-long supply of nitrogen. Adequate soil moisture in the presence of adequate plant nutrition will assist in maintaining overall plant health

— Potatoes should be planted after risk of frost has passed and when rapid emergence will reduce risks of seed decay

Pest management considerations

— The main threat to yield loss is from potato beetle .

— Bacillus thuringiensis (Bt), a biological insecticide is accepted by many certification agencies, but producers are advised to consult with their own agency as to which Bt strains are approved for use

— Isolation from production areas with a high density of potato acres will reduce risks from all insect pests

— Locating a potato field at least 200 - 300 meters from a previous season's field will reduce the population of over-wintering potato beetle adults and resulting larvae. In spite of the success of this strategy be prepared with a plan to control larvae.

- Weed control
- In extreme cases, summer fallow the year prior to potato production may be required to reduce weed populations to a point where they will not reduce yields or interfere with harvest operations
- Post-plant cultivation (hilling, harrowing and hilling) is effective in controlling annual weeds, however, excessive cultivation or cultivation at the wrong time may reduce yield

Effect of Diseases

- Low-generation certified seed must be used to reduce the risk of seed-borne diseases
- Whole seed must be used to reduce risk of spreading disease during cutting
- Most whole seed pieces will emerge and produce healthy plants. However, some cut seed pieces will always decay resulting in reduced final stand. Plants will grow from seed tubers that are partially decayed, however vigor and yield will be reduced
- Isolation may reduce the risks from diseases such as late blight.

References

Brook, R. C., R. J. Fick, and T. D. Forbush, *Potato Storage Design and Management*, East Lansing, MI.: Michigan State University Agricultural Engineering Information Series, 1995.

Christ, B. J., *Potato Diseases in Pennsylvania*, Pennsylvania College of Agricultural Sciences, University Park, PA. 1998.

Dunn, J. W., J. K. Harper, and G. L. Greaser, *Agricultural Alternatives: Fruit and Vegetable Marketing for Small-scale and Part-time Growers*, University Park, PA: Penn State Cooperative Extension, 2000.

Greaser, G. L. and J. K. Harper, *Agricultural Alternatives: Enterprise Budget Analysis*, University Park, PA: Penn State Cooperative Extension, 1994.

Lamont, W. J., *et al.*, *Agricultural Alternatives: Irrigation for Fruit and Vegetable Production*, University Park, PA: Penn State Cooperative Extension, 2001.

9

Irrigation Management

Knowledge about the amount of soil water available to the plant and the ability to estimate soil water content is a key aspect in determining irrigation requirements of potatoes. An irrigator must understand the following terms to estimate soil water content.

Available soil water (ASW) is a common term of reference describing soil water content. Understanding the concept of ASW requires understanding of two additional terms related to soil water content. Field capacity refers to the amount of water a soil holds after water has drained out of the larger pores and moved out through the soil profile due to the force of gravity.

In coarse textured soils it may take a few hours for a soil to move from a saturated status to field capacity, in medium textured soils this may take approximately 24 hours, while in fine textured soils this may take several days.

Permanent wilting point refers to the amount of water a soil holds when only water within the smallest of the pores and spaces around individual soil particles remains. This water is not considered available to plants since the energy required to extract it is too great for the plant roots.

Any water held in the soil pores between the soil water contents of field capacity and permanent wilting point is considered available for extraction by plant roots for growth.

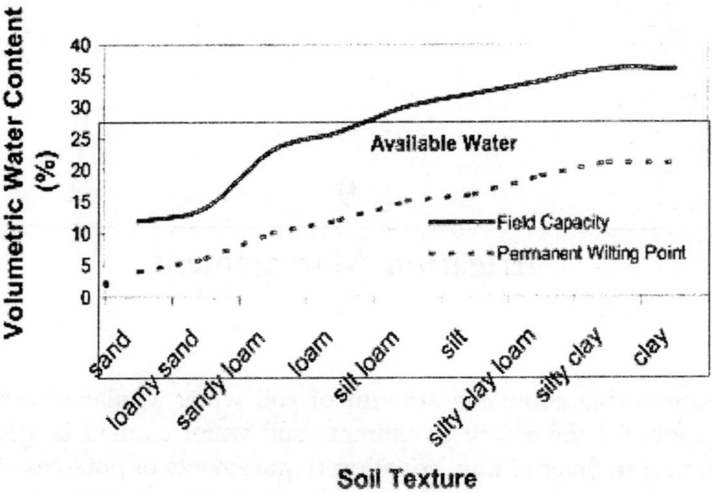

Figure 1. Available Soil Water and Soil Texture

The ASW capacity is the amount of water represented by the difference between field capacity and permanent wilting point. Soil texture is the main soil characteristic affecting ASW (Figure 1), however other factors such as structure, salinity, mineralogy and organic matter content also play a role.

Characteristics of Crop

Potato is a moisture sensitive crop with a shallow active root zone compared to cereals and forages. Roots will penetrate to 3 feet (90 cm), however about 90% of the root activity takes place in the top 2 feet (60 cm) depth with a significant amount of that activity in the upper foot (30 cm). Availability of moisture in the root zone is crucial for high yields and is influenced by soil properties such as texture and percent organic matter. To maximise potato yield, maintain the soil moisture in the root zone above 65% of the ASW capacity. The optimum temperature for potato growth is 69° F (21° C).

Severe growth restrictions occur when the temperature is below 45° F (70° C) or exceeds 86° F (30° C). This information is used to calculate heat units for potatoes, which are called P-Days. Figure 2 charts the rooting depth and growth stages of a Russet Burbank crop based on accumulation of P-Days. Root depth is important when estimating the amount of soil water available to the plant. A grower with access to P-Days for their production area can use the model in Figure 2 to estimate rooting depth and in turn the amount of soil water available to the potato plant.

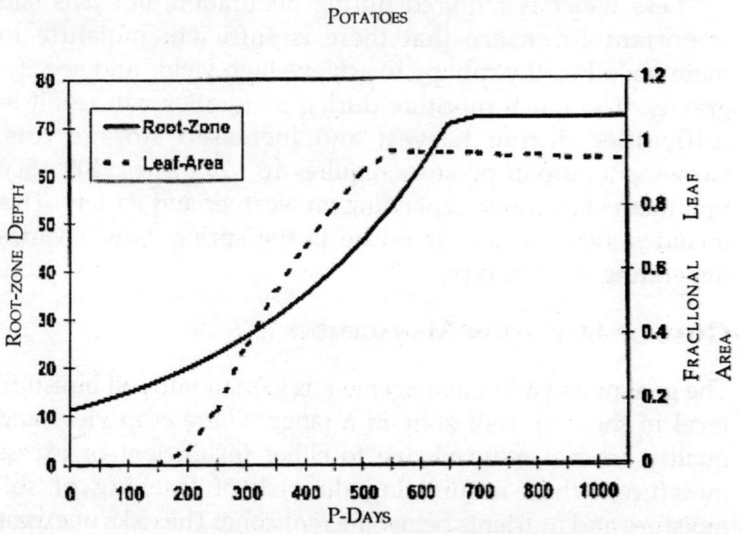

Figure 2. Potato growth model for a Russet Burbank crop grown in Manitoba

The growth and development of potatoes can be broken down into five distinct stages, which are discussed in detail in Growth Stages. Moisture demand increases as the crop begins to develop after emergence and peaks 7-9 weeks later during the tuber bulking stage of growth. Demand begins to decrease from the time of late bulking to maturation. To achieve

maximum yield and quality, soil moisture should be maintained at or above 65% of ASW capacity for all stages of growth.

If soil moisture content is allowed to drop below 70% ASW during the tuber set/initiation, a reduction in tuber number per plant will occur. Water stress at any time during tuber bulking will reduce marketable yield. Low water levels or excessive fluctuation of water levels outside the desirable range can also reduce quality, contributing to growth deformities such as hollow heart, knobbiness and growth cracks.

Less water is required during maturation, but it is still important to ensure that there is sufficient moisture to maintain a healthy canopy to achieve high yields and specific gravity. Too much moisture during maturation can result in difficulties during harvest and increased storage rots. Growing a crop of potatoes requires 16 - 20 inches (400 - 500 millimetres) of water depending on weather and variety. This includes available soil moisture in the spring, which varies depending on soil type.

Goal of Irrigation Management

The goal of irrigation management is to maintain soil moisture level in the crop root zone in a range where crop yield and quality are not reduced due to either insufficient or excess moisture, while minimising the risk of leaching of soil moisture and nutrients below the root zone. The risks of excess moisture and leaching are greater in Manitoba than in many other potato-growing areas, due to heavier growing season precipitation.

Effective rooting depth (Figure 2) must be considered when assessing soil moisture available to the crop for irrigation scheduling purposes. A depth of two feet (60 cm) is often assumed for management of irrigation scheduling for potatoes; this would be applicable for most of the irrigation season.

It is generally recommended that irrigation be scheduled to prevent root zone soil moisture from dropping below 65% ASW. That critical ASW for irrigation is somewhat higher for fine-textured soils than for sands, since moisture is held more tightly in finer soils at a given ASW.

The amount of water a soil can hold between field capacity (100% ASW) and 65% ASW varies from about one-third inch per foot of soil (2.8 cm per m) for very coarse soils, to about two-thirds inch per foot (5.6 cm per m) for loam to finer textured soils. The amount of irrigation water applied with each application should be set to not exceed this capacity of the soil to hold water within the desirable range.

Otherwise, leaching through the root zone will occur. For example, if water is to be applied at 65% ASW, and a 24" (60cm) effective rooting zone is assumed, routine application of 1.25" (3.2 cm) of water per irrigation would be acceptable on a clay loam soil, but only up to about half that amount on a very coarse soil.

Therefore applications should be smaller and more frequent on coarse soils. If soil texture varies considerably within the field, the maximum application amount should be based on the capacity of the coarsest textured soil that occurs over a substantial portion of the field.

Methods of Irrigation Scheduling

Most irrigation scheduling methods are based on either information from direct determination of soil moisture in the field (feel method), or on soil water balance calculated from estimated crop water use (checkbook method), or on a combination of the two approaches. Variability in soil characteristics across the field, as well as with depth will result in uneven soil moisture content and must be considered in design and use of any irrigation scheduling system. In addition to soil characteristics and soil water holding capacity, other factors will result in variations in moisture availability across the field. These include:

— Variability in crop vigour and growth

— Plant spacing
— Slope, which moves rain and irrigation water off of the field before it can infiltrate into the soil
— Hilling, which can also affect run-off and infiltration
— Non-uniformity of irrigation water application.

A balance must be struck between the costs of under-irrigating the driest parts of the field and those of over-irrigating the wettest portions. Experience, which includes years of checking soil moisture levels throughout the field under various conditions, is invaluable. Scheduling based on actual soil moisture determination should ideally incorporate many sampling/monitoring points within each field. With fewer points used, the grower should scout more frequently to verify that those points are representative of most of the field.

Scheduling based on water budgeting methods should always include some field scouting for verification. The amount of scouting required increases with moisture variability over the field.

The most basic approach for determining the amount of soil water involves sampling of the soil in the root zone and estimating moisture content by the "feel and appearance" method.

Many soil samples representing the different soil types in the field must be collected to obtain an accurate estimate of moisture content. Though this method is laborious and requires experience to interpret the soil moisture content, it can be an effective way of estimating the ASW.

A wide range of instruments is commercially available to measure soil moisture for irrigation scheduling. These include neutron probes, tensiometers and resistance blocks. Some indicate how tightly the water is held by the soil.

Many indicate the amount of water in the soil, usually by volume. For both, it is necessary to understand the relationship between soil water tension and content for the specific soil to interpret the results for best irrigation management. Most of the instruments are installed in the soil

at the beginning of the irrigation season, eliminating the need to obtain soil samples.

Some can even be equipped to transmit readings by telemetry. Water balance methods of scheduling are based on calculated soil moisture levels through the growing season.

All inputs and removals of water from the root-zone are measured or estimated frequently (usually daily), and a running "balance" of soil moisture is calculated. At the start of the season, soil moisture level is either determined, or assumed to be at field capacity.

Soil water inputs are rainfall (less estimated runoff) and irrigation. Soil water removals are evapotranspiration (ET) and deep drainage through the root zone, which is estimated from inputs and soil water holding capacity. Evapotranspiration is a combination of moisture loss from the soil surface by evaporation and from the plant tissue by transpiration.

The simplest approach is to estimate daily ET based only on crop age or stage of growth and daily maximum temperature. More accuracy can be obtained using more frequently measured atmospheric condition, such as solar radiation, wind, temperature, and humidity.

Other approaches use a device from which water evaporates under the local conditions. Daily ET is then calculated from the amount of water evaporated, by application of a factor for the crop stage.

A computer is usually used to do the daily calculations of soil water levels from the water input/output data. Irrigation is used to keep the soil moisture in the desired range.

Since the checkbook method uses only an estimate of ET, and relies upon several assumptions about water behaviour in the soil, the actual soil moisture should be checked periodically and corrected. Under normal soil and weather conditions, checks are recommended every two weeks.

Suitability of Land for Irrigated Potato Production

The suitability of soil conditions and landscape features for

irrigated potato production is based on soil resource information from detailed soil maps (1:20000 to 1:50000 scale) and soil databases.

Information considered in evaluating soil characteristics and landscape features are:

- Soil texture and thickness and uniformity of deposits
- Topography
- Stoniness
- Salinity
- Soil drainage
- Soil Order and Subgroup

Only the dominant soil series and landscape features for each polygon are used in the evaluation and in map production. Subdominant soil conditions and landscape features should also be considered when assessing a given land area for irrigated potato production.

The rating of soil and landscape characteristics for a given land area will assist in making an initial assessment of land suitability for irrigated potato production.

Description of Suitability Classes

Class 1 - Lands in this group generally have the following characteristics:

- soils are well to rapidly drained
- soil textures usually range from loamy very fine sand, sandy loam to loam
- soils are nonstony and have level to very gently sloping (0 to 5 %) topography
- salinity levels in the soil profile (0 ~60 cm) are very low

Class 2 - Lands in this group generally have the following characteristics:

- soils are well to rapidly drained
- soil materials usually include sandy (coarse) textures such medium and fine sand, loamy sand; and finer textures such as clay loam and silty clay loam

— soils are nonstony and have level to very gently sloping (0 to 5 %) topography

— salinity levels in the soil profile (0 ~60 cm) are very low

Class 3 - Lands in this group generally have the following characteristics:

— soils are well to rapidly drained

— salinity levels in the soil profile (0 ~60 cm) range from 0 to 4 mS/cm (nonsaline)

— soil materials usually include sandy (coarse) textures such medium and fine sand, and loamy sand underlain by silty clay and clay materials; and finer textures such as loam, clay loam and silty clay loam underlain by coarse sand and gravel

— soils are nonstony and have level to very gently sloping (0 to 5%) topography

— soils are imperfectly drained

— textures are similar to soils in class 1 and 2

— salinity levels in the soil profile (0 ~60 cm) range from 0 to 4 mS/cm(nonsaline)

— soils are nonstony and have level to very gently sloping (0 to 5%) topography

Class 4 - Lands in this group generally have the following characteristics:

— soils are well to imperfectly drained

— salinity levels in the soil profile (0 ~60 cm) range from 4 to 8 mS/cm(weakly saline)

— soils are slightly stony and have gently sloping topography (5 to 9 %)

— soil materials usually include a range of textures underlain by clay textured materials

Class 5 - Lands in this group generally may have one or more of the following characteristics:

— salinity levels in the soil profile (0 ~60 cm) are greater than 8 mS/cm(moderately to strongly saline)

— soils are moderately to excessively stony (st. 2,3,4,5) and have moderate to very steep slopes (greater than 9 %)
— soil textures are usually gravels, and/or clay; soils underlain by bedrock within 120 cm and soils with textural complexes, undifferentiated and fragmental materials
— organic, gleysolic and solonetzic soils

REFERENCES

Baek Sauk Gi., *Jang Young Sil.* Woong-Jin-Wee-In-Jun-Gi: "Woongjin Publishing Co., Ltd"., 1987.

Dillehay TD, Eling HH Jr, Rossen J., "Preceramic irrigation canals in the Peruvian Andes", *Proceedings of the National Academy of Sciences*, 2005.

Frenken, K., *Irrigation in Africa in figures – AQUASTAT Survey – 2005*, Food and Agriculture Organization of the United Nations, March 2007, <ftp://ftp.fao.org/agl/aglw/docs/wr29_eng.pdf>.

Needham, Joseph, *Science and Civilization in China: Volume 4, Physics and Physical Technology, Part 2, Mechanical Engineering*, Taipei: Caves Books Ltd., 1986.

Rodda, J. C. and Ubertini, Lucio, *The Basis of Civilization - Water Science?* pg 161, International Association of Hydrological Sciences, International Association of Hydrological Sciences Press 2004.

Snyder, R. L. & Melo-Abreu, J. P., *Frost protection: fundamentals, practice, and economics – Volume 1*, Food and Agriculture Organization of the United Nations, 2005, <ftp://ftp.fao.org/docrep/fao/008/y7223e/y7223e00.pdf>.

Siebert, S.; J. Hoogeveen, P. Döll, J-M. Faurès, S. Feick, and K. Frenken, "The Digital Global Map of Irrigation Areas – Development and Validation of Map Version 4", *Tropentag 2006 – Conference on International Agricultural Research for Development*, 2007-03-14.

Williams, J. F.; S. R. Roberts, J. E. Hill, S. C. Scardaci, and G. Tibbits, Managing Water for Weed Control in Rice, UC Davis, Department of Plant Sciences, 2007-03-14.

10

Harvest Management

The majority of the cost of producing a potato crop has accrued by the time of harvest and growers should not risk that expenditure and effort by an untimely harvest. The factors that should be considered when planning a timely harvest are:

— Target yield

— Crop maturity

— Soil and tuber temperatures

— Harvest capacity

Ideally, harvest would not begin until the crop has attained the producer's target yield goals, but that is not always achievable for late maturing varieties. The growing season in Western Canada is highly variable and there are not always enough heat units to achieve target yield goals.

Delaying harvest will allow for additional growth and yield, but any gains in yield may be offset by losses, if not all of the crop can be harvested and stored because of adverse weather (rain or frost). Chemical maturity contributes to storability, marketability and processabilty of the tubers.

Delaying harvest will usually improve maturity, but any gains may be offset by losses, if not all of the crop can be harvested and stored because of adverse weather (rain or frost). Crop maturity monitoring does not affect the onset of harvest, just the order which fields are harvested. The grower should harvest the most mature fields first leaving the

immature fields to the end of harvest, allowing for more growth.

Harvest temperatures have the greatest influence on harvest timing. The ideal harvest temperature is between 45 and 60°F (7° and 15°C). Tubers warmer than 64-68°F (18-20°C) and under drought stress are susceptible to black spot bruising. Harvesting when tuber pulp temperature exceeds 68° F (20°C) increases the risk of leak and pink rot, which can result in storage decay.

Harvesting when the tuber temperature is less than 45°F (7°C) will contribute to the incidence of shatter bruising. If the tubers are repeatedly chilled below 40°F (5°C) the concentration of sucrose will be high resulting in poor fry colour. Depending upon hill shape, variety, soil moisture and vine cover; temperatures below 32°F (0°C) will freeze some or all of the tubers, causing the crop to break down in storage. Rarely can a crop with more than five percent frozen tubers be stored successfully. To bring a good quality crop into storage, the harvest must take place when soil and tuber temperatures are ideal.

The number of days that these temperatures exist in September and early October are limited. The grower must have the capacity to complete harvest before chilling temperatures or frost damage will render the crop unmarketable. In early October, the risk of crop damage is very high and producers should plan to complete harvest before then. When evaluating harvest capacity, the grower must account for some down time due to inclement weather, wet soils and mechanical breakdowns.

Chemical Maturity of Potato Tubers

The 'chemical' maturity of potato tubers provides a measure of their suitability for long-term storage and processing quality. As potato tubers develop and the crop matures, the amount of free sugars (sucrose, glucose, and fructose) in the tuber decrease. When the sugars drop to a minimum level, the tubers are considered "chemically mature". This stage

normally coincides with maximum dry matter accumulation and indicates that the crop is ready for harvest. Late planting that shortens the growing season, cool weather that delays plant growth, excess nitrogen, and heat stress can cause elevated sugars at harvest.

Chemically immature potatoes contain a high concentration of sucrose, which breaks down to produce the reducing sugars glucose and fructose after the tubers are harvested. High reducing sugars in turn result in unacceptable darkening during frying. Storage stresses such as improper handling, cold temperatures, high carbon dioxide, improper sprout inhibitor application, and tuber aging can also increase sugars, especially in tubers, which were chemically immature at harvest. Chemical maturity is determined by the sugar concentration in the tuber.

Analysis of the sucrose and glucose concentrations along with a standard frying test provides useful information in assessing the chemical maturity of the crop and the effect of various management practices on crop quality. Sucrose concentration gives a good estimate of potential processing quality before harvest while glucose concentration, when measured during storage, gives the best estimate of processing quality of French fry cultivators, specifically Russet Burbank and Shepody.

Sucrose monitoring can be used to determine which fields are nearing maturity and glucose monitoring can be used to determine when they have reached a stage of chemical maturity that will ensure optimum processing quality out of long term storage. Different varieties vary in the amount of sugar that is acceptable for processing. For example, chipping potato varieties must have a lower reducing sugar content than those used for French fries. Determination of sugar concentrations can be accomplished by various technologies. Most processors use some method of sugar analysis. Equipment for this purpose can be expensive although less expensive technologies are available though may be less reliable.

Method of Vine Killing

Timely vine killing is essential in the production of table, chipping and seed potatoes. Vine killing is not always employed in the production of late season maturing French fry processing potatoes. The growing season in Manitoba is short and late season tuber growth is required to achieve an economic yield of processing potatoes, so vine killing is rarely used. The growing season in Alberta is longer, economic yields are achieved early in the season, so vine killing is a more common practice. Vine killing achieves the following:

- — Termination of growth to control tuber size
- — Termination of growth to set skin, which reduces skinning and bruising during harvest and shrinkage in storage. A requirement for table production
- — Prevents the spread of viral diseases by aphids.
- — Reduction in the spread of late blight spores from infected vines to tubers at harvest
- — The time between vine killing and harvest allows diseased tubers to rot in the hill before harvest
- — Stabilisation of chemical maturity in chipping potatoes
- — Reduction in harvest losses by facilitating separation of the tubers from the stolen
- — Removal of vine growth that interferes with harvest

If potatoes are not fully mature, vine killing can reduce the yield and specific gravity. Manitoba studies report that vigorous Russet Burbank potatoes develop 5.5 to 8.6 cwt/ac/day (620 - 970 kg/ha/day) in early September. Vine kill during this stage of growth can significantly reduce yield. The extent of yield and specific gravity losses depend upon the following factors:

- — Stage of growth at the time of vine killing. The more immature the crop at the time of vine killing, the greater the loss of yield and specific gravity
- — Condition of the crop at the time of vine killing. Yield losses are greatest if the crop is vigorous and capable of further tuber bulking

— The interval between vine killing and harvest or a killing frost. The longer the interval between vine killing and harvest or frost, the greater the yield and specific gravity losses

— The effectiveness of the vine killing operation. Mechanical vine killing will terminate growth much more rapidly than chemical desiccation and result in a greater yield loss if carried out too early

Under rare circumstances, discolouration of the vascular ring can occur after vine killing. Discolouration is confined to the stem end of the tuber, causing problems in processing (French fries and chips) and table markets. Stem end discoloration (SED) has no effect on seed quality. SED is more likely to occur when vine killing is rapid, the soil moisture is low and the weather is hot. To reduce the risk of SED, avoid rapid vine killing methods during hot weather, especially if the soil is dry.

Vine killing is accomplished by chemical desiccation and mechanical methods. Chemical desiccation is the most common method of vine killing. Apply the desiccant 2-3 weeks before harvest to allow for complete vine kill. Plants may continue to bulk after desiccation, so apply in advance of achieving the desired tuber size for seed and table production. Variety, plant vigour and environmental conditions will effect the rate of desiccation. Two chemical desiccants are registered for use in Western Canada.

Chemical Desiccation

Reglone produces a moderate rate of vine kill. Two or 3 applications may be required if growth is vigorous. Reglone performs best if applied at dusk with high volumes of water. This allows the product to penetrate the leaves and stems before being activated by the sunlight the following day.

As well, best results are obtained after the onset of senescence and adequate skin set has been established. Liberty produces a slow to moderate rate of vine kill. A single application will usually achieve complete vine kill. Liberty

works best when applied during daylight hours, when photosynthesis is taking place. Efficacy is also enhanced when application is delayed until after the onset of plant senescence. Liberty works best when warm and humid. Conditions that impede plant growth will reduce the effectiveness of Liberty. Liberty is not registered for use on seed potatoes.

Mechanical Methods of Vine Killing

Two methods of mechanical vine killing are employed on the Prairies. Both are used in conjunction with a chemical desiccant. Rolling vines with rubber tires or crowfoot packer wheels followed by a Reglone or Liberty application produces a moderate to fast rate of vine kill. Rolling damages the foliage resulting in partial vine kill, but it also exposes more of the foliage to Reglone or Liberty, enhancing the effect of the desiccant. This results in fewer desiccant applications to achieve complete kill and a faster rate of vine kill. The vine roller also seals cracks in the hills, preventing sunlight from contacting the tubers and causing greening. The flail vine shredder, which consists of varying knife lengths to accommodate the contours of the hill and furrow, produces a very rapid vine kill.

Flailing removes all but 6 inches (15 cm) of the vine. Flailing closer to the hill can uncover and injure the tubers. The debris from flailing can interfere with the performance of Reglone. A short waiting period may be required for the debris to fall off of the remaining vine before applying Reglone. Do not use the vine shredder in conjunction with Liberty.

Some flail implements are equipped with drum rollers, which seal cracks in the hills, preventing sunlight from contacting the tubers and causing greening. Flailing reduces the desiccant applications required to achieve complete vine kill and provides a significantly faster rate of vine kill. Flailing causes rapid death, so the vine kill procedure can be timed closer to the harvest date than slower vine kill methods. This extends the length of time for tuber bulking and chemical maturation, thus maximising yield and specific gravity.

Types of Potato Bruising

An excellent crop in the field prior to harvest does not ensure a high profit out-of-storage. A severe frost or excessive precipitation during harvest, which is beyond the control of the grower, can devastate a crop and eliminate any profit. Harvest timing and management, which can be controlled by the grower, also has a significant effect on quality, net yield and profit out-of-storage. Mechanical damage and bruises caused by rough handling during harvest, trucking and piling into storage, will increase the incidence of storage rots, reduce quality, increase grade-out and reduce the value of the crop. There are two types of potato bruising that occur at harvest. Both types are caused when the potato strikes an object, however, each type of bruise occurs under contrasting environmental conditions.

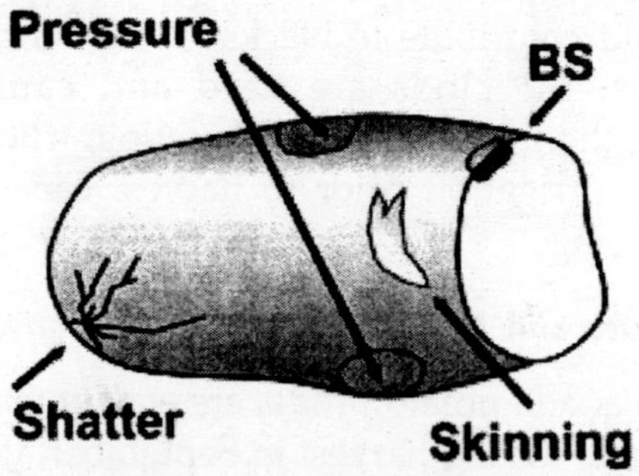

Figure 1. Blackspot and Shatter bruise

Blackspot bruising occurs under warm dry soil conditions. Susceptibility increases with increasing tuber temperature and decreasing soil moisture and water content in the tuber. The impact ruptures individual cells below the skin of the tuber without breaking the skin. These bruises are not immediately visible. After two days the damaged tissue will turn dark grey or black and can be seen only after the skin is peeled. Varieties vary greatly in their susceptibility. Shatter bruises are thin cracks or splits in the tuber flesh. This is more

common on larger tubers. Susceptibility increases with decreasing temperatures below 45°F (7°C) and increasing soil moisture and water content in the tuber.

Thumbnail cracks are a form of shatter bruise, which can occur when cold tubers are roughly handled out of storage or on packing lines. Other forms of mechanical damage include skinning, cutting and scraping. All types of bruises and mechanical damage adversely affect the appearance of potatoes and can result in rot in storage. The amount of bruising and mechanical damage at harvest is influenced by soil conditions, cultivar, tuber maturity, tuber temperature, and equipment condition and operation at harvest. For detailed harvester adjustments, please contact your local machinery dealer.

Production of Soil Clods

Field preparation or in-row tillage performed under wet soil conditions can produce soil clods. Soil clods will persist until harvest and contribute to blackspot bruising. Under dry soil conditions, the clods are hard and cause injury when contacting the tubers during harvesting, trucking and storing operations. Irrigation prior to harvest can soften clods and reduce damage.

Temperature and Moisture

Western Canada potato producers encounter a wide range of temperatures during harvest in September and early October. Extremes in temperature cause different types of bruising. Hot and dry soils produce ideal conditions for blackspot bruising. Irrigating before harvest will increase the water content of the tuber decreasing susceptibility to blackspot bruising. The risk of shatter bruise increases later in the season as soil temperature decreases. Growers should complete harvest early to avoid conditions, which cause this type of bruising. Harvest date is influenced by factors such as cultivar selection, planting date, fertility, irrigation, and vine killing.

Fertility

Excess nitrogen delays maturity and promotes skinning and blackspot bruising.

Harvester Operation

The harvest operation accounts for the majority of mechanical damage to potatoes. Trucking, piling into storage and handling for market also contribute to bruising. The following points should be considered when preparing and operating the harvester and other handling equipment:

- All chain or conveyor links and flights should be properly padded with the exception of the primary bed
- Use ample padding on deflectors, sharp edges and any other locations where bruising can occur
- Adjust the digger blade depth so that potatoes are not be bruised or cut. Operate the blade deep enough to maintain some soil on the primary and secondary bed to cushion the potatoes
- Adjust the digger blade height on harvesters and windrowers so potatoes do not bump into the front of the primary chain
- Maintain a proper forward speed to chain ratio to ensure good soil separation and, at the same time, keeping the conveyors as full of potatoes as possible
- Keep chain agitation to a minimum. Consider installing hydrostatic agitators (shakers) on the primary bed to improved control
- Keep beds and conveyors sufficiently tight to avoid sagging and whipping
- Reduce the distance that tubers drop to 15 cm or less including the loading boom. Consider installing deflector lip on the end of the boom to reduce bruising
- Check frequently for bruising at various points on both harvester and equipment

Make your employees aware of the causes and consequences of bruising. Train them to operate equipment in a manner to

prevent damage to the tubers. Encourage operators to inspect all handling equipment regularly and to inform you of any situations that have the potential to cause bruising. Encourage employees to act as a team to prevent bruising. Since harvest operators cannot see the entire digging operation they must rely on the knowledge and experience of others for feedback.

References

Food and Agriculture Organization of the United Nations, *Prevention of postharvest food losses: A training manual*, Rome: FAO Training Series, 1985.

Kader, A.A., ed., *Post-harvest technology of horticultural crops*, Oakland: University of California, Division of Agriculture and Natural Resources Publication 3311, 2002.

Kitinoja, L. and A.A. Kader, *Small-scale post-harvest handling practices: A manual for horticultural crops*, Fourth edition. Davis: University of California, Postharvest Horticulture Series 8E, 2002.

Mitra, S., ed., *Post-harvest physiology and storage of tropical and subtropical fruits*, Wallingford, UK: CAB International, 1997.

Shewfelt, R.L. and S.E. Prussia, eds., *Post-harvest handling: A systems approach*, San Diego, CA: Academic Press, 1993.

Thompson, A.K., "Post-harvest technology of fruits and vegetables", Oxford: Blackwell Science, 1996.

Thompson, J.F., A.A. Kader, and K. Sylva, *Compatibility chart for fruits and vegetables in short-term transport or storage*, Oakland: Univ. Calif. Div. Agric. Nat. Res. Publ. 21560 (poster), 1996.

Wills, R., B. McGlasson, D. Graham, and D. Joyce, *Postharvest: An introduction to the physiology and handling of fruit, vegetables and ornamentals*, Wallingford: CAB International, 1998.

11

Agrometeorology and Potato Production

The potato (*Solanum tuberosum L.*) is a major world food crop and by far the most important vegetable crop in terms of quantities produced and consumed worldwide. Potato is exceeded only by wheat (*Triticum aestivum L.*), rice (*Oryza sativa L.*), and maize (*Zea mays L.*) in world production for human consumption. Potato tubers give an exceptionally high yield per acre and are used in a wide variety of table, processed, livestock feed, and industrial uses. Potato provides nutritious food in a diversity of environments. Potato can be an important food for the increasing world population, and has the potential for increased vitamin C and protein content.

The mainly limiting factors for potato production are heat and water stresses. The effects of these factors on physiology, yield and grade of potato crop are thoroughly discussed in the current contribution. The meteorological elements governing growth, development, production, and quality of potato tubers at a given site are basically air and soil temperatures, solar radiation, photoperiod, soil moisture, and crop water use or evapotranspiration.

The aboveground stems of potato plants are erect in early stages of development but later become spreading and prostrate or semi-prostrate. The tuber is an enlarged underground stem. The tubers have buds or eyes, from which sprouts arise under certain conditions. Tubers are harvested for both food and seed. The flowers and fruits are only important to potato breeders.

Potato has a relatively shallow, fibrous root system with the majority of the roots in the surface 0.3 m. The root system develops rapidly during early growth and achieves maximum development by mid season. Thereafter, root length, density, and root mass decrease as the plant matures. Rooting depths of 1.2 m or more have been reported for potato under favourable soil conditions. Potato extracted less water from the soil than barley (*Hordeum vulgare L.*) and sugarbeet (*Beta vulgaris L.*) and the differences were accentuated below 0.6 m depth.

The origin of potato in cool climates with equatorial day lengths, and the shallow potato root systems have consequences for the agrometeorological responses of the crop. Knowledge of climatic requirements of potato and its physiological responses to the environment is extremely important to help growers produce high yields with good tuber quality under site-specific atmospheric conditions. Crop weather models can be used to provide estimates of potato yield as a function of climatic factors at a particular locality. The SUBSTOR-Potato model, for instance, takes into consideration daily data of temperature, photoperiod, intercepted solar radiation, soil water and nitrogen supply. The model simulated fresh tuber yields ranging from 4 Mg ha-1 to 56 Mg ha-1 due to differences in climate, soils, cultivars and management practices.

Potato plant has five growth stages: sprout development (I), plant establishment (II), tuber initiation (III), tuber bulking (IV), and tuber maturation (V). Timing and duration of these growth stages depend upon environmental factors, such as elevation and temperature, soil, moisture availability, cultivar and geographic location.

At growth stage I, sprouts develop from eyes on seed tubers and grow upward to emerge from the soil, roots begin to develop at the base of emerging sprouts, and the seed piece is the sole energy source for growth during this stage. At stage II, leaves and branches develop on emerged sprouts; roots and stolons develop below ground, and photosynthesis begins.

Potato development in stages I and II lasts from 30 to 70 days, depending on planting date, physiological age of the seed tubers, cultivar, soil temperature, and other environmental factors. At stage III, tubers form at stolon tips but are not yet appreciably enlarged and in most cultivars the end of this stage coincides with early flowering with an average duration of roughly two weeks. At stage IV, tuber cells expand with the accumulation of water, nutrients, and carbohydrates. During tuber bulking stage, tubers become the dominant site for carbohydrate and inorganic nutrient storage. Tuber bulking can continue up to three months as a function of the cultivar and environmental conditions. During stage V photosynthesis gradually decreases, leaves turn yellow, tuber growth rate slows, and the vines die. Maturation may not occur in the field when a long season variety like Russet Burbank is grown in a short season production area.

Agroclimatology of the Crop

There are three phenological phases in the allocation of dailyaccumulated dry matter. Initially, dry matter is divided between stems and leaves (growth stage II). In the second phase, which starts at tuber initiation, an increasing amount of accumulated dry matter is allocated to the tubers and a decreasing fraction to the leaves (growth stages III and IV). In the third phase all assimilates are allocated to the tubers (growth stage V). Leaf growth stops and photosynthesis eventually stops because of leaf senescence. Climatic factors influence all three phenological phases. The duration of the first phase, comprising the development period between emergence and tuber initiation, is shortened by short days and temperatures less than 20°C. Tuber initiation is slower at temperatures over 20°C. The duration of the second phase is affected by temperature with an optimum between 16 and 18°C or 14 and 22°C and by solar radiation. Crop senescence is shortened by high temperatures, especially greater than 30°C. The effects of agroclimatological factors on physiological parameters of potato, especially on tuber yield, grade, and internal quality, will be discussed below.

Air Temperature, Solar Radiation and Photoperiod

Due to the interactive effects of air temperature, photoperiod (day-length), solar radiation, and cultivar on the tuberisation stimulus, these meteorological variables will be discussed together with emphasis on physiological responses to one or another climatic element consistent with the specific objectives of each research project. The review by Haverkort points out that potato is best adapted to cool climates such as tropical highlands with mean daily temperatures between 15 and 18°C as encountered in its center of origin. Higher temperatures favor foliar development and retard tuberisation. In addition, heat stress leads to a higher number of smaller tubers per plant, lower tuber specific gravity with reduced dry matter content, and usually to a paler skin color of the tubers.

Temmerman et al. examined the effect of latitude, seasonal mean air temperature (ranging from 13.8 to 19.9°C), global solar radiation (ranging from 12.0 to 21.3 MJ m^{-2} d^{-1}), air humidity, soil moisture, and atmospheric CO_2 concentrations on tuber yield in European experiments. Ignoring CO_2 enrichment, the yield of potato increased from south to north Europe. Marketable tuber yields increased at higher latitudes. The authors ascribed this result to lower temperatures, lower vapor pressure deficits, and longer day lengths at higher latitudes, which in turn resulted in longer effective growing seasons.

Climatic conditions, as affected not only by the latitude but also by altitude, influence potato plant growth and development. Moreno found that plants grown at low (coastal) altitudes have low yield of tubers per plant as compared with those grown in the Andean highlands. Tubers harvested from coastally grown plants had lower free amid acid and amide contents and a higher content of tuber protein than those from the Andean highland. Coastal tubers also had less total sugar content than Andean tubers.

Haverkort reports that an inconvenience of the short day sensitivity of the potato is that cultivars that make use of the whole growing season and produce well in northern Europe

(5-6 month growing season), may mature too early and senesce between 60 and 70 days after planting in the equatorial highlands and consequently yield less. Cultivars that perform well at low latitudes in a 3 to 4 month growing season start tuberising late and mature too late at 50°N.

Photoperiodic responses are mediated by endogenous plant hormones. Relatively high gibberellic acid (GA) levels reduce or stop tuber growth and relatively high abscisic acid (ABA) levels promote tuber growth. In some potato cultivars and species, long photoperiods produce high GA levels that prevent tuber growth. This can be a problem for temperate regions, which have long photoperiods during their usual crop season. Fortunately, many of the American cultivars are "day neutral" and presumably have lost the GA-photoperiod response.

Carbon dioxide concentration can also exert a strong influence on potato productivity. The influence of carbon dioxide depends on solar irradiance. Potato (cultivars 'Norland', Russet Burbank, and 'Denali') were grown at CO_2 levels of 350 or 1000 mmol mol^{-1}, irradiance of 400 or 800 mE m^{-2} s^{-1} photosynthetic photon flux (PPF), and photoperiod of 12 or 24 hours-light. Increased CO_2 provided greater tuber yield at low PPF but decreased tuber yields at high PPF. Increasing the PPF increased the tuber yield for Denali but decreased the yield for Russet Burbank. When averaged across all irradiance treatments, Denali showed the greatest gain in tuber and total weight (21 and 18%, respectively) in response to increased CO_2 enrichment for the three cultivars tested. Norland showed the least (9 and 9%, respectively), while Russet Burbank showed an intermediate response, with gains nearly as great as for Denali under a 12 h-photoperiod (18%) but less than Denali under a 24 h-photoperiod. A pattern of greater potato plant growth was observed from CO_2 enrichment under lower PPF and a short photoperiod.

Crop growing systems for space travel are needed to generate oxygen, purify water, remove carbon dioxide, produce food, and recycle waste materials. Total irradiance has been suggested to be the largest limitation to crop

productivity in these systems. Potato yield improvements might be obtained by increasing the net daily photosynthetically active radiation (PAR) through higher irradiance or longer photoperiod.

The photoperiod duration doubles from December to June at 50°N, while PAR increases eightfold from 211 to 1701 MJ m^{-2} d^{-1} due to higher elevation of the sun above the horizon with lengthening days. Gross carbohydrate production on standard clear days increases from 108 to 529 kg ha^{-1} d^{-1} at 50°N, whereas it remains at about 420 kg ha^{-1} d^{-1} year-round near the equator. Low solar irradiance is a yield constraint at 30 to 40°N in winter when potatoes are grown to escape the summer heat.

Stuttle et al. studied the effect of photoperiod (12, 18, and 24 h-light) on net carbon assimilation rate (A_{net}) and starch accumulation in newly mature canopy leaves of Norland potato under low and high PPF, 263 and 412 mE m^{-2} s^{-1}, respectively. Whenever the photoperiod was increased from 12 to 18 hours, there was a marked decline in A_{net} of 16.1%, and declines were most pronounced under high PPF. The maximum starch concentrations were obtained under high PPF treatments at a shorter photoperiod than under low light treatments.

An apparent feedback mechanism exists for regulating A_{net} under high PPF, high CO_2, and long photoperiod, but there was no correlation between A_{net} and starch concentration in individual leaves. This suggests that maximum A_{net} cannot be sustained with elevated CO_2 enrichments under long photoperiod and high PPF conditions for Norland. Therefore, if a physiological limit exists for the fixation and transport of carbon, increasing photoperiod and light intensity under high CO_2 enrichment may not maximise potato yield. Since the onset and early phases of tuber growth are important for the further development of potato,

Dam et al. conducted a factorial experiment with two photoperiods (12 or 18 h) and four 12-h day/night temperatures to analyse photoperiod and temperature effects

on early tuber growth, drymatter partitioning, and tuber number for cultivars 'Spunta' and 'Desiree'. They concluded that low mean temperatures (15-19°C) with a short photoperiod (12 h) were most suitable for early tuber growth. Under these conditions, onset of growth and onset of bulking were early, and absolute tuber growth rates and dry matter partitioning were high.

Slight increases in temperature strongly reduced partitioning rates, whereas further increases had a large impact on the onset of tuber growth and absolute growth rates. Differences between treatments in numbers of tubers initiated were inconsistent. The absolute growth rate under long photoperiod was higher for Spunta than for Desiree. Different genotype responses to temperature and photoperiod on tuber growth were also found by Snyder and Ewing using potato cuttings.

Midmore and Prange examined the effects of day/night temperature (33/25°C or 20/10°C), and 12-h high irradiance (430-450 mE $m^{-2} s^{-1}$ PAR) or 12-h low irradiance (250-280 mE $m^{-2} s^{-1}$ PAR) both with a 6-h photoperiod extension at 6 mE $m^{-2} s^{-1}$ on relative growth rate, net assimilation rate, and dry matter production of Solanum goniocalyx cv. 'Garhuash Huayro' and DTO-33, a heat tolerant clone of S. tuberosum x S. phureja. The highest relative growth rate was obtained at low temperature and low irradiance.

At high temperature, low irradiance had the opposite effect, producing the lowest net assimilation and relative growth rates. Both tuber number and weight were markedly reduced by high temperature. Low irradiance in combination with high temperature produced virtually no tubers. These data, consistent with field observations that reduced potato growth at high temperatures, can be aggravated by lower irradiance. Both leaf area and net assimilation rate are reduced.

Manrique and Bartholomew carried out a potato genotype x environment experiment on Mt. Haleakala, Maui, Hawaii, at three elevations from 91 to 1097 m, to assess the

performance of four standard temperate cultivars and three heat-tolerant clones in warm to cool temperatures at photoperiods prevailing in the tropics. Dry weight of plant components and total dry weight per plant were measured at tuber initiation, 20 days after tuber initiation, and 40 days after tuber initiation.

Warm temperatures at 91 m hastened development such that, at tuber initiation, total dry weight per plant was 2 to 4 times greater than at 1097 m in 1985 and 1986 Tuber dry weight increased significantly at the second two sampling dates with lower temperature at higher elevation. Dry matter partitioning to tubers generally was highly and significantly correlated with temperature, with the optimum of 15 to 20°C for tuber growth. Potato plants lost their ability to allocate dry matter to tubers at higher temperatures.

Sarquis et al. stated that the magnitude of the effect of elevated temperatures on potato growth and final yield is determined by an intricate interaction between soil temperature, air temperature, solar radiation and photoperiod duration. Their data extended previous observations of reduction in photosynthesis rate under elevated temperatures. Under field conditions they concluded that reduced carbon assimilation rate could not explain the yield reduction observed; the temperature effect on assimilation was not as dramatic as it was on growth or yield. Other workers have reported a severe reduction in the rate of assimilation at air temperatures above 30°C under controlled experimental conditions. In such cases, the reduction in carbon assimilation rate was shown to correlate well with reductions in growth and yield. These contrasting results reveal the complexity of plant responses to the combined effects of water and temperature stress, which inevitably occur in association under field conditions.

Thornton et al. examined the effect of two day/night air temperature regimes (low 25/12°C, and high 35/25°C) on dry matter production of three potato clones for five weeks, beginning two weeks after tuberisation, under controlled

environmental conditions. Tuber growth rate was more affected by high temperature than was whole plant growth. All clones exhibited a decline in tuber dry matter production at high compared to low temperatures; however, Russet Burbank exhibited the largest decline. Potato clones varied in partitioning of dry matter to tubers at high temperatures. In addition to carbon assimilation, heat stress reduced tuber yields by affecting several plant processes such as dark respiration. Although high temperature stress is a major uncontrolled factor affecting growth, development and productivity of plants, relatively little is known about genetic diversity for heat tolerance in potatoes. Tolerance to heat stress may involve many complex relationships. An adapted genotype must have a diverse and complex combination of genes for tolerance to high temperatures and for superior performance in the field.

Potato cultivars and clones vary significantly in their ability to tuberise at elevated air temperatures and continuous irradiance. Tibbitts et al. carried out two experiments under controlled environments to determine the capability of 24 highly productive potato genotypes to tolerate continuous light and high temperature. Six cultivars grew well under continuous light while three cultivars were superior to the others at high temperature. Two cultivars were well adapted to continuous light and high temperature. These evaluations were made after only 56 days of growth and further assessments should be made in longer-term productivity studies. For some crop plants, leaf angle can be important for maximising solar radiation interception.

With potato cultivars that are intercepting as much as 95% of incident solar radiation at a LAI of 4, one must question whether alterations in leaf angle would significantly improve light interception. Individual leaves can utilise only 50-60% of incident radiation on a clear day. Following tuber initiation, the photosynthetic apparatus saturates by about 1200 mE $m^{-2}s^{-1}$, or about 60% of full light. Ideally, the top leaves of a potato canopy should absorb no more than 1200 mE $m^{-2}s^{-1}$ and should allow the remaining light to pass to the lower canopy.

Opportunities remain to modify potato plant architecture to increase productivity.

Gawronska and Dwelle studied the effect of high light levels (maxima between 500 and 1200 mE m^{-2} s^{-1}) and shaded low light levels (approximately one-quarter of the high light) on potato plant growth, biomass accumulation and its distribution. They observed that plants under low light did not produce auxillary shoots, while those under high light did. Tubers of plants under low light were very small and irregular in shape. The most evident plant response to low light was greater stem elongation as well as a reduction in total biomass accumulation and in tuber weights. The reduction in total biomass under low light was 34 to 45%. Reduction in tuber dry weights under low light ranged from 39 to 57%, depending on the growth stage and harvest time. In addition, at all growth stages, the percentage of biomass partitioned to the tubers was higher under high light than under low light conditions.

According to Gawronska et al. potato plants grown under low light generally had lower rates of photosynthesis (when compared with those grown under high light), reaching saturation for maximum photosynthesis at about 500 mE m^{-2} s^{-1}. Some clones maintained the higher rates of photosynthesis than Russet Burbank at nearly all-light levels, demonstrating the potential to breed for cultivars that maintain higher rates of photosynthesis and potentially higher tuber yields.

Soil Temperature and Soil Temperature Management

The rate of development of sprouts from planted seed pieces depends on soil temperature. Very little sprout elongation occurs at 6°C. Elongation is slow at 9°C and is maximised at about 18°C. The time between planting and emergence depends on soil temperature. Phytotron and field experiments carried out by Sale showed that emergence was linearly related to mean soil temperature and relatively independent of diurnal fluctuations up to an optimum of 22-24°C. Up to this optimum emergence could be considered as a degree-day

requirement calculated either from soil temperature at tuber depth or air temperature. At temperatures above the optimum, emergence was inhibited.

Sattelmacher et al. studied the effect of 20°C and 30°C root-zone temperatures on root growth and root morphology of six potato clones. Significant genotypical differences in the responses of potato roots to 30°C were observed, indicating the potential for selecting heat tolerant potato clones. In both heat tolerant and heat sensitive clones, the size of the root system was reduced by a 30°C root-zone temperature explained by a reduction in the cell division followed by cessation of root elongation.

Tuberisation stimulus favors both tuber initiation and tuber enlargement. Through artificially prolonged exposure to short days and cool temperatures, it is possible to attain such a high level of stimulus that induction is irreversible, even if potato plants are subsequently exposed to long days for weeks or months. The optimum soil temperature for initiating tubers ranges from 16 to 19°C. Reynolds and Ewing examined the influence of four air and soil day-night temperature treatments on root, tuber, and shoot growth in growth chambers: (cool air (19/17°C), with cool or heated soil (20/18°C or 32/31°C); and hot air (34/30°C), with hot or cooled soil (32/27°C or 19/17°C)). Cooling the soil at high air temperatures neither relieved visible symptoms of heat stress on shoot growth nor increased the degree of induction tuberisation by the leaves.

Heating the soil at cool air temperatures had no apparent detrimental effect on shoot growth or induction of tuberisation by the leaves. Under high soil temperatures, stolonisation was substantially compromised and there was no underground tuber development. In one experiment, stolons grew up out of the hot soil and formed aerial tubers above the soil surface in the cool air. The induction of tuberisation by the leaves was affected mainly by air rather than soil temperature, but the signal to tuberise might be blocked by high soil temperatures.

According to Mares et al., it is expected that the effect of high soil temperature on growing tubers would be similar to that of exogenously applied gibberellin, inhibiting tuberisation. Tuber development declines as soil temperatures rise above 20°C and tuber growth practically stops at soil temperatures above 30°C. The number of tubers set per plant is greater at lower temperatures than at higher temperatures, whereas higher temperatures favor development of large tubers. Little research is available on the effect of soil temperature during tuber growth on potato grade and quality. Kincaid et al., assessing the influence of the interaction between water management and soil temperature on potato quality in the Pacific Northwest, observed that the critical period for tuber quality appears to be from mid-June to mid-July, based on measured soil temperature differences, frequent sprinkler irrigation reduced soil temperatures, along with the incidence of sugar-end tubers.

Yamaguchi et al. found that yield, specific gravity and starch content of Russet Burbank and 'White Rose' tubers were higher, and the sugar content lower when grown at soil temperatures between 15 and 24°C, than when grown at higher temperatures. Ewing reports that in many areas the sequence of temperatures that most often brings economic damage to potato crops is warm temperatures early in the season, followed by cool temperatures that induce strong tuberisation, followed in turn by another period of high temperatures. Such temperature oscillations lead to heat sprouts, chain tubers, and secondary growth of tubers. Apparently the fluctuations in tuberisation stimulus cause tuber formation to alternate with more stolon-like growth.

Management practices, such as planting population density, use of mulch and irrigation might substantially modify the soil temperature regime within the root zone in such a way as to affect stolonisation, tuber initiation and bulking, and tuber enlargement at a given site, particularly where solar irradiance availability is shown to be a nonlimiting factor for potato production. Increase of plant population through a reduction of between-row spacing was

effective in raising tuber yields in the hot tropics, largely through the increase in amounts of intercepted solar radiation, which brought about a significant decline on soil temperatures during the tuber growth. Since the proportion of marketable tubers was scarcely affected by planting densities, Midmore reasoned that potato plant population in hot climates should be as high as possible without limiting the amount of soil available for hilling-up.

In order to quantify the effects of organic mulch on soil temperature and soil moisture regimes during the growth of potato, Midmore et al. conducted seven experiments at three contrasting hot tropical sites (latitude varying from 5 to 12°S, and altitude ranging from 180 to 800 m). Mulch retained more heat in the soil at night when combined with agronomic practices that themselves increased soil heat retention at night (i.e. on the flat potato beds). The magnitude of soil cooling by mulch during the day and heat retention within the soil at night was dependent on solar irradiance levels and soil moisture content. Mulch was more effective in cooling dry soils, especially at high irradiance.

Heat retention at night following days of low irradiance was greater in mulched plots, whereas at high irradiance heat retention of mulched plots was intermediate between those of moist and drier control plots. Midmore et al. showed that mulch increased tuber yield by 20% during the summer in Lima, Peru. Manrique and Meyer, studying the impact of mulches on potato production during winter and summer seasons at the same site, found no effect on yields during the winter, but yield increases of 58% and improvements in soil moisture retention were obtained in the summer with surface mulch.

Mahmood et al. reported that mulch at Islamabad, Pakistan, decreased daily maximum soil temperature at a 15 cm-depth by 1.5 to 4.5°C, resulting in faster emergence, earlier canopy development, and higher tuber yields. Many other recent studies conducted in Asia point out the beneficial effects of mulch in potato production systems as an efficient

alternative to obviate heat and water stresses in order to maximise crop yield.

Atmospheric Humidity and Wind, Wind Management

There are very few recent studies dealing with the direct effects of relative humidity (RH) on potato growth, tuber yield and grade. Most of the contributions related to the influence of RH on potato refer to potato storage where RH is an important factor in tuber weight loss and the occurrence and severity of diseases and pests. The same scarcity of research exists with regard to the wind regimes at a particular location as an agrometeorological factor affecting potato production systems. Wheeler et al. studied the effect of two RH levels, 50% and 85%, on the physiological responses of three cultivars of potato in controlled-environment rooms under continuous light intensity at 20°C. No significant differences in total plant dry weight were measured between the atmospheric humidity treatments, but plants grown under 85% RH produced higher tuber yields.

Leaf areas were greater under 50% RH and leaves tended to be larger and darker green under drier than at more humid atmospheric conditions. The elevated humidity appeared to shift the allocation pattern of photosynthates to favor allocation to the tubers over leaves and stems.

Gordon et al. estimated sap flow from solar radiation and vapor pressure deficit data for three field-grown potato cultivars ('Atlantic', 'Monona' and 'Norchip') at Nova Scotia, Canada, under non-limiting soil water conditions. Sap flow rates for all cultivars were closely linked with solar radiation under conditions where soil water was not limiting. The vapor pressure deficit (VPD), a function of relative humidity and air temperature, had less effect on sap flow, although the magnitude of the VPD during the growing season was generally < 2 kPa. All cultivars maintained actual daily transpiration near the potential energy limiting rate under well-watered conditions. When the soil was drier (percent available soil water < 30%), Monona potato plants had a much

more rapid decline in transpiration than the other two cultivars.

Another physiological parameter closely related to yield is water use efficiency. Bowen reported that potato farming in coastal Peru occurs during the winter, when the cool humid conditions favor growth and promote a more efficient use of irrigation water. During the winter, less soil water evaporation caused by a smaller VPD enhances water use efficiency when compared with that observed during the summer. Sinclair also showed that generally more humid environments provide greater water use efficiency because of a lower VPD.

Stomatal resistance governs photosynthesis and transpiration. Two major feedback loops are reported by Raschke as the direct controllers of stomatal resistance (rst). The first involves photosynthesis where a reduction in intercellular carbon dioxide (CO_2) occurs as the photosynthetic active radiation (PAR) increases, the stomata open and rst decreases. The second involves an increase in rst whenever leaf water potential reaches a critical threshold as a result of transpiration intensity. Stomatal resistance is affected by many factors including PAR, the ratio of leaf to air water potential, leaf age, air temperature and the ambient CO_2 concentration.

Gordon et al. studied the stomatal resistance of three field grown potato cultivars (Atlantic, Monona and Norchip) in response to photosynthetic photon flux density, leaf to air vapor pressure difference and root zone available water. Under the climatic conditions of their field experiment in Eastern Canada, stomatal activity in potato was primarily driven by light intensity. However, as soil water became limiting the soil/plant water status became increasingly more important. The absence of very high VPD values throughout the growing season is the probable main reason for the lack of potato rst response to air vapor pressure differences. Significant differences were observed among cultivars in the response of stomata to changes in available soil water.

Crop weather modeling needs to incorporate these differences into model systems because they might have a significant effect on eventual model performance at a given site. Wind has important effects on potato. Pavlista reported that leaves injured by lower wind speeds show bronzed areas, brown with a shiny surface, due to the rubbing of leaves against each other. The bronzed areas tend to brittle from drying. When pressed the bronzed areas crack, forming a sharp-edged rip through the affected tissue. Under higher wind speeds, leaves not only bronze but also tatter. Tattered leaves typically have a 6 to 25 mm sized tears with irregular brownish borders.

Stems may also be affected by winds. When exposed to a mild wind, stems may just be flopped around causing a slight weakness of the tissues. Under strong winds, vines might actually get twisted, bringing about a break or hinge-like weakness in the stems. If exposed to strong winds for several hours, the vine may twist all the way around and cause the stem to collapse, cutting off nutrient flow through the phloem between the vine and the tubers. Wind also affects transpiration rates and, therefore, photosynthetic activity and crop yield. At sites where winds are frequently strong throughout the year, increased stomatal resistance can cause reduction in potato yield. At such sites, guidelines for the sustainable management of potato cropping systems need an emphasis on windbreak development including height, porosity, and orientation.

Sun and Dickinson studied the benefit of two 30-month-old windbreaks (one with two rows of trees and one with three rows of trees) for potato in tropical northeastern Australia. Two Eucalyptus species (E. microcorys and E. torelliana) were found to be highly suitable for windbreaks since they showed rapid development in height and branch growth while retaining low branches. The porosity of three row and two row windbreaks were 37.2 and 60%, respectively. The optimum range of porosity for windbreaks is between 40 and 50%. Windbreaks increased potato plant growth in height and leaf

number, however, had limited effects on leaf length and width.

Potato plants grown close to windbreaks yielded more than those grown at the furthest positions, with the highest production removed 3 times the windbreak height. Windbreaks increased potato yield by up to 7.7%, whereas Sturrock found windbreaks increased yield by 35%.

Wright and Brooks examined the effect of windbreaks on growth and yield of potatoes over a 4-year period in Australia, measuring the amount and severity of wind damage to leaves, plant height, and leaf numbers from potato located at various distances from the windbreak in both sheltered and unsheltered positions. Windbreaks increased tuber yield between 4.8 and 9.3% for the sheltered portion of the field in seasons with higher than average wind speeds and caused a reduction in wind damage to leaves on protected potato plants. In seasons where wind speed was above average, windbreaks increased yield at distances away from the windbreak between 3 and 18 times the height of the windbreak. Cleugh reported that potato crop yields were significantly higher in the sheltered zone from 2 to 18 times the height of the windbreak compared to yields obtained in unprotected areas.

Crop Evapotranspiration and Irrigation Requirements

Crop consumptive water use is the amount of water transpired by the plants plus the water evaporated from the soil plus the fraction of water held by the plant tissues. The amount of water retained by plant metabolic activity is about 1% of the overall water taken up by the plants. Thus, in practical terms crop water consumption corresponds to crop evapotranspiration (ET_c). Potato ET_c can be estimated using weather data and is the amount of water to be replenished during the growing season in order to assure potential tuber yields at a given site. Potato ET_c is important to consider in irrigation planning and its use in irrigation scheduling is a well-developed strategy to improve the effectiveness of irrigation.

An adequate water supply is required from tuber initiation up until near maturity for high yield and good quality. Applying water in excess of plant needs compromises the environment, may harm the crop, and is expensive for growers. Excessive irrigation of potatoes results in water loss and significantly increases of runoff and soil erosion from production fields to rivers, streams, and reservoirs. Leaching can lead to contamination of the groundwater due to lixiviation of fertilisers and other chemical products.

Irrigation in excess of crop needs increases production costs, can reduce yield by affecting soil aeration and root system respiration, and favors the occurrence and severity of diseases and pests. Deficient irrigation promotes a reduction of tuber quality and lower yield due to reduced leaf area and/ or reduced photosynthesis per unit leaf area. Local atmospheric conditions, surface soil wetness, stage of growth, and the amount of crop cover are the factors that govern the daily fluctuations of potato ET_c. They observed that ET_c increased as the leaf area and transpiration increased and reached near-maximum levels just before effective full cover.

The leaf area index (LAI) reached 3.5 by effective full cover coincident with the highest daily ET_c of 8.5 mm. Seasonal total ET_c corresponded to 604 mm in southern Idaho. Potato ET_c varies greatly from region to region. Seasonal potato ETc in the humid Wisconsin area for June through August ranged from 293 to 405 mm during 3 years of study. At Mesa, AZ, ET_c for February through June, averaged 617 mm. The mid season daily potato ET_c was 6 mm near Calgary, Alberta, Canada, while the daily water consumption was 3 mm under the climatic conditions of Botucatu, State of São Paulo, Brazil, during the winter where seasonal ET_c was only 283 mm.

Wright and Stark reported that seasonal water use in irrigated areas of Oregon and Washington, USA, ranged from 640 to 700 mm. For high yields at a given site, the seasonal water requirements of a potato crop with a phenological cycle varying from 120 to 150 days ranged from 500 to 700 mm, depending on climate. The maximum daily potato ET_c

measured by a weighing lysimeter in a sub-humid region in India was found to be 4.24 mm d^{-1}. Under a hot and dry climate in northeastern Portugal, peak ET_c rates reached 12-13 mm d-1 on the days immediately following irrigation, but crop water use declined logarithmically with time to about 3 mm d^{-1} within 5 days.

Wright developed improved crop coefficients for various irrigated crops in the Pacific Northwest, including potato, using alfalfa to measure reference evapotranspiration (ET_o) and weighing lysimeters at an experimental field near Kimberly, Idaho. Growth-stage specific crop coefficients (K_c) and the water balance method provided a valuable tool in scheduling overhead irrigation of Russet Burbank potatoes in the Columbia Basin of Oregon. Simonne et al. reported that K_c values ranged from 0.3 at emergence to 0.8 during maximum leaf area, and declined as the crop matured. ET_c is usually calculated by the product of K_c and ET_o, or as a function of a number of climatic elements to provide the atmospheric potential demand.

Apart from the crop coefficient approach, potato evapotranspiration can also be estimated by means of multiple regression equations that take into consideration the leaf area index (LAI) of potato crop and atmospheric evaporative demand depicted by ET_o or pan evaporation.

Potato can be sensitive to irrigation less than ET_c that result in soil water deficits. A study in three successive years on silt loam soil in eastern Oregon investigated the effect of water deficit on yield and quality of four potato cultivars grown under four season-long sprinkler irrigation treatments. The results suggest that irrigation water applied at rates less than ET_c in the Treasure Valley of Oregon would not be a viable management tool to economise water, because the small financial benefit would not offset the high risk of reduced tuber yield and profit from the reduced water application.

Potato cultivars may respond differently not only to deficit irrigation but also to total seasonal crop

evapotranspiration under non-limiting soil water supply. Wolfe et al. reported that total seasonal actual crop water use at Davis, California, on a deep Yolo loam soil ranged from 316 to 610 mm for the 'Kennebec' cultivar and from 331 to 630 mm for 'White Rose', as a function of six levels of irrigation water supply established throughout the growing season. Shock et al., comparing the performance of two new potato cultivars with four other cultivars grown in the Treasure Valley of Oregon, observed that Umatilla Russet showed a higher yield potential at ideal water application rates, while Russet Legend was the only cultivar tolerant to deficit irrigation treatments.

ET_c is an essential agrometerological index, which can be used to determine both the amount of water to be applied and the irrigation frequency for a particular crop and site. Stockle and Hiller compared a canopy temperature-based method, the neutron probe method, and the computer-assisted method based on evapotranspiration and K_c values to schedule irrigation for potato in central Washington State. A soil water depletion of 70% was allowed before starting irrigation. They concluded that the most practical method was the computer-assisted method using estimates of ET_0 and K_c values.

Soil Moisture Requirements and Irrigation Management

Soil moisture status is expressed by percent available soil water (ASW) content or by soil water tension (SWT). Available soil water content is defined as being the amount of water that plants can extract from a given volume of soil, from the crop effective rooting zone. Available soil water is usually expressed as a percent between "field capacity" (100%) and "permanent wilting point" (0%). Soil water tension is the force necessary for roots to extract water from the soil.

Curwen reviewed water management for potato and placed great emphasis on using the irrigation criterion of 65% ASW. At "field capacity" (100% ASW), the SWT is often between 20 and 33 kPa depending on soil type and the method of determination. Soil water is assumed to no longer be available at the "permanent wilting point", generally assumed to be at a SWT of 1,500 kPa. The ASW approach works well

for irrigation scheduling in regions with extensive areas of homogeneous soil. It is often a practical impossibility for growers to know when the soil is at 65% ASW, even if they have soil water content sensors available.

Usually the percent water content that a given field contains at "field capacity" is unknown for a given part of a specific field. Similarly the percent water content at the "permanent wilting point" for a given part of a specific field is also usually unknown. Both the "field capacity" and the "permanent wilting point" vary tremendously with soil type, vary from spot to spot within a field, vary with cultivation, and vary over time. With neither "field capacity" nor "permanent wilting point" known, 65% ASW cannot be known; the prescription of an irrigation criterion of 65% ASW can become much like telling a grower to irrigate at the "right moment" and leaving the decision to intuition and experience.

Growers need direct and unambiguous irrigation recommendations to deal with crops that have negative responses to small variations in irrigation management. In contrast to the ASW, SWT can be measured directly using tensiometers or granular matrix sensors. The SWT irrigation criterion needed to optimise potato yield and quality can be determined by production region and generalised soil type. Measurements of SWT that optimise potato yield and grade have been determined for a number of locations, some of which are wetter than 65% ASW.

Based on potato yield and grade responses to irrigation, ideal potato SWT irrigation criteria were found to be 50 kPa using furrow irrigation on loam in California, 50 to 60 kPa using sprinklers on silt loam in Oregon, 25 kPa using sprinklers on silt loam in Maine, 60 kPa and 30 kPa using furrow and drip irrigation, respectively, for silt loam in Oregon, and 20 kPa using sprinklers on sandy loam in Western Australia.

Irrigation Scheduling

Irrigation of crops sensitive to water stress requires a systematic approach to irrigation scheduling. Information to

answer irrigation scheduling questions may include atmospherically based, plant-based, or soil-based data. Examples of atmospheric irrigation scheduling information include weather forecasts, estimates of crop evapotranspiration (ET_c) such as those provided by AgriMet, pan evaporation, and atmometers. AgriMet is an automated weather station network operating throughout the Pacific Northwest of the USA that uses site-specific climatic data, the current stage of growth of local crops, and models to estimate daily crop water use.

Plant data may include canopy temperature, xylem water potential, and visible wilting. Soil-based data may include soil water content and soil water tension (SWT). In practice, plant, soil, and atmospheric data are often used concurrently, especially when changes in irrigation schedules are required to adjust for changes in crop water use. Growers should pay attention to crop appearance, soil water tension, the rate of crop evapotranspiration, precipitation, and the amount of water applied. With knowledge of these factors, irrigation can be well managed in order to obtain high yields of better quality tubers along with environmental protection.

Other Information on the Crop Yield and Quality

Response to Irrigation Management

Potato tuber response to soil moisture conditions begins before tuber set. MacKerron and Jefferies have shown that increased duration of water stress before tuber initiation reduces tuber set per stem. Shock et al. demonstrated that reduced tuber set in the Treasure Valley was related to the duration of SWT drier than 60 kPa before and during the beginning of tuber set. Where Verticillium wilt is present, there are advantages to keeping soils a little dry early in the season before tuber initiation.

Jones and Johnson described the reduction in potato yield caused by water stress. Through the use of a line source sprinkler system, Hang and Miller showed how a moisture gradient affects plant top growth, tuber yield, and tuber

grade. With sprinkler irrigation, water application had to remain near potential ET_c for maximum tuber yield and grade.

Potato varieties differ in their response to water stress. Kleinkopf found that Russet Burbank was more sensitive than the 'Butte' variety in forming misshapen tubers under water stress.

Assuring Tuber Grade

Fluctuations in water that stress the potato plant during tuber development can result in greater proportions of misshapen tubers of lower market grade. Corey and Myers determined that the proportion of misshapen tubers was directly related to drier SWT. Eldredge et al. found that a single transient SWT stress drier than 50 kPa increased misshapen Russet Burbank tubers. Pereira and Villa Nova studied the effect of three irrigation treatments on tuber yield and grade at Botucatu, São Paulo, Brazil. Potatoes irrigated to fully replace ET_c had higher yields and better grade and fewer physiological defects.

Assuring Internal Tuber Quality

Tuber physiological disorders such as brown center, hollow heart, and translucent end, as well as secondary growth, growth cracks, bruise susceptibility, and heat necrosis have been associated with water stress and/or wide variations in soil moisture content.

The sugar-end disorder is also known as dark ends, translucent ends, or in more severe incidences when stem-end tissue breakdown occurs, jelly ends. Jelly ends can occur in the field or during storage. These physiological disorders are often considered a minor production problem. However, when above normal temperatures occur during the growing season, significant economic losses can result from excess reducing sugars (glucose and fructose) in the stem end of tubers. These reducing sugars react with free amino acids during frying to form brown or black colors. For processors dark-ends result in reduced processing efficiency and economics and in some cases, an unusable product.

When dark ends, as measured at harvest, exceed contract specifications, grower returns are reduced by contract penalty clauses. Research has shown that the incidence of sugar ends in tubers was reduced substantially when irrigation scheduling was based on SWT measurements. Dark-ends may become more severe after tubers have been stored. The timing of water stress is important; water stress before tuber initiation has no deleterious effect on tuber quality, while stress later during tuber bulking can cause dark stemend fry color and reduced specific gravity.

Penman was one of the first authors to discuss the importance of translucent-end potatoes. Numerous authors have suggested that translucent-end potatoes are caused by early season moisture stress. Sugar-end tubers result in French fry "dark-ends" and are related to tubers with translucent-ends and jelly-ends. Owings et al. reproduced the results of Iritani and Weller demonstrating that late June water stress could cause sugar ends. Shock et al. subjected potatoes to water stress in May and the beginning of June and found very early season stress did not result in sugar-ends. But short duration water stress any time during tuber bulking accompanied by heat stress resulted in sugar ends.

Increases in reducing sugars occurred more than two weeks after the end of transient water stress, which suggests that water stress causes enzymatic or membrane changes that eventually result in the loss of cellular control of sugar metabolism and the onset of sugar-ends and translucent ends. Sowokinos et al. demonstrated the importance of specific tuber starch and sugar enzymes in the development of sugar-ends.

Paradoxically, season-long uniform stress does not have the same negative effect on potato tubers. Painter et al. observed no fry color differences between potato irrigated at 25 ASW season-long and those irrigated at 65 ASW, which proved to be consistent with later findings where potato was stressed all season. Water stress is often associated with increased canopy temperature and soil heating in the field. In

most field trials where water stress has been imposed and measured, canopy and soil temperatures have not been measured.

Irrigation and Microclimate Interaction with Potato Diseases and Pests

Irrigation management practices can affect disease severity. The increased humidity from irrigation will have greater effects where the macroclimate is humid or sub-humid and be of less importance where it is drier. For potato grown in hot areas, sprinkler irrigation can cool the environment, with possible reductions in physiological defects. However, different irrigation methods can contribute to the occurrence of diseases and pests on the crop depending on site-specific weather pattern. Wet soil is conducive to most tuber-rotting pathogens. Excessive soil moisture following planting can promote seed piece decay and erratic plant. Excess soil moisture also encourages the incidence of blights, rots, and wilts, particularly prolonged excess soil water conditions.

Avoiding over-irrigation, or even keeping soils a little dry early in the season before tuber initiation may reduce the amount of root infection by V. dahliae, a major component in early die. On the other hand, avoiding excessive plant water stress during the tuber bulking growth stage, which usually coincides with the warmest part of the season, may help decrease the severity of early die. Potato vines that remain wet for long periods create a microenvironment conducive to early blight (Alternaria solani), late blight (Phytophthora infestans), white mold (Sclerotinia sclerotiorum), and blackleg (Rhizoctonia solani). The timing of these diseases and associated crop losses vary regionally with yearly weather patterns, and can be affected by irrigation methods, which increase or decrease the duration of high humidity in the crop canopy.

Consistently rainy summer or fall weather promotes late blight. However, in the 1990's, epidemics of late blight developed in potato crops in arid production areas of the Pacific Northwest where late blight had not been a problem.

Irrigation that tends to keep the foliage wet may contribute to this developing risk. Potatoes cultivated under center pivot irrigation can receive a relatively low volume of irrigation water for a long time near the pivot, favoring late blight occurrence. Johnson et al. showed that the incidence of late blight tuber rot significantly increased as the amount of irrigation water applied increased, and was significantly greater within 30 m of the pivot than at greater distances. Long duration sprinkler irrigation also favored late blight in Oregon and California.

Long periods of leaf wetness or high relative humidity within the potato canopy favor infection by white mold. Avoiding light, frequent irrigation of coarse-textured soils, and avoiding heavy, less frequent irrigation of fine-textured soils can diminish the risk of white mold. Simons and Gilligan found irrigation to increase the incidence of stem canker, stolen canker, and black scurf to a limited extent although the effect of season tended to be more pronounced on these defects than any of the agronomic treatments tested.

While avoiding developing high humidity in the canopy, adequate soil moisture is essential not only for potato yield and quality but also for pest management strategies. Adequate soil moisture helps reduce the attack of cutworms (Spodoptera litura) and mites (*Tetranychus spp and Tenuipalpidaee spp*). Potato tubermoth (Phthorimaea operculella) and its larvae are repelled by soil moisture. Soil moisture also reduces formation of cracks in the soil, which allow the entry of potato tubermoth and its larvae. Irrigation scheduling based on ET_c and/or SWT can account for the local climate and keep the soil from becoming too dry.

References

Cappaert, M.R., M.L. Powelson, N.W. Christensen, W.R. Stevenson, and D.I. Rouse, "Assessment of irrigation as a method of managing potato early dying", *Phytopathology*, 1994.

Feustel, I.C., "Miscellaneous products from potatoes", *In* Talburt, W. F., O. Smith (eds.) *Potato Processing*, 4th Ed., Van Nostrand, New York, 1987.

12

Potato Storage Management

Almost any type of building can be adapted to store potatoes, however, commercial rigid frame steel buildings are not normally used because the exposed steel beams and columns are difficult to insulate. The most common storage buildings are concrete, wood stud and pole frame, and metal quonset.

The factors that vary between various building types are capital cost, durability and longevity, and the type of insulation required for the exterior building envelope. Regardless of the type of building, the design should be undertaken by a Professional Engineer to ensure that the structure can withstand the forces exerted by the stored potatoes, wind and snow. Design and construction can typically require three to four months or more; therefore early planning is required to ensure the storage is ready at harvest time. The size of individual bins and storage systems vary widely with the needs of individual producers and the cost of construction.

Bin sizes in modern buildings normally range from 40,000 to 80,000 cwt. Potatoes from individual fields often behave differently in storage, requiring individual management. Ideally, the potatoes from each field should be stored in a separate bin, but since smaller bins cost more per hundredweight, storage management is compromised when storing potatoes from different fields in large bins. An enclosed loading area attached to the storage bins is recommended. This will allow workers to comfortably load

potatoes in severe weather and minimise potential problems associated with chilled tubers.

Effect of Insulation

Storages must be properly insulated and sealed in order to maintain the environment required to keep stored potatoes healthy. Besides reducing heat loss and thus helping to maintain the desired storage temperature, insulation is also critical in preventing condensation. Condensation water, dripping onto the tubers, will encourage the development of soft rots and can significantly impact potato quality. It is recommended that enough insulation be installed to achieve a minimum thermal resistance (RSI) of 6.1 (R35). This is equivalent to 10 inches (250 mm) fibreglass or 6 inches (150 mm) of polyurethane insulation.

Ceiling fans have also proved beneficial in reducing free moisture on ceiling surfaces and/or the top of the potato pile. The amount of insulation also impacts interior air quality. Insulation decreases heat loss through the walls and ceilings, resulting in more heat of respiration exhausted from the building via the ventilation system. In extremely cold weather, this allows the ventilation system to bring additional fresh air into the building, thus maintaining adequate levels of oxygen and reducing the level of deleterious gasses such as carbon dioxide.

Ventilation System

Ventilation is the most important factor for maintaining correct temperature, relative humidity, and air quality in the storage. It is also essential for managing potential storage problems caused by disease or frost. The basic ventilation system design is similar, regardless of the type of storage structure. A typical ventilation system consists of intake doors, fans, air plenums, ducts, exhaust louvers and a control system. A qualified individual should design the ventilation system.

The size of the intake doors, air plenums, ducts, duct outlets and exhaust louvres must be carefully selected to ensure that the ventilation air is evenly distributed throughout the storage. Intake dampers should be designed to close return air supply proportionally as the intake door is opened. As more fresh air intake is required, return air carrying excessive heat or humidity will be forced out of the building through the exhaust vents. This can be particularly helpful when trying to remove excessive field heat at harvest.

The addition of refrigeration coils, humidification units and light traps in the exhaust louver all impact the resistance to airflow and must be considered when selecting the ventilation fans. Controllers for varying the speed and airflow of the fans are recommended. Contact Manitoba Agriculture or a potato processor for the name of someone qualified to design the ventilation system. Most control systems utilise a single insulated damper to control the blend of fresh and return air. A heating system, in the perimeter of the damper, is used to prevent freesing of the damper in cold weather. Heavy-duty screw type actuators are used to adjust the position of the damper.

Ventilation controllers vary in complexity, depending on the number of control strategies. The simplest strategy involves running the fans continuously. The volume of air is manually adjusted through the number of fans operating or by adjusting the speed of the fans. In this situation, the minimum sensors required are: the temperature control sensor that modulates the mixing damper; a low limit in the supply air to prevent accidental chilling or freezing of the tubers; and an outdoor sensor to prevent outdoor air warmer than the pile from entering the storage. More advanced systems have carbon dioxide (CO_2) sensors that will activate the damper to bring in fresh air whenever CO_2 levels exceed a preset limit. Few ventilation control systems include a humidistat because they are inaccurate when the relative humidity is >90% and have been of limited value controlling humidity in a potato storage. Under normal storage

conditions, the relative humidity of the supply air should be maintained near 98 percent.

Humidifiers should be installed immediately downstream from the fan(s). Three types of humidifiers are commonly used: high-pressure nozzles, centrifugal spinning disk, and water-saturated fibrous media. The first two types are used less frequently as they are difficult to regulate, resulting in either too much or not enough water added to the air streams. The third type humidifies the air as it passes through a fibrons media without the preasure of free water droplets that can effect potato quality. This type of humidification unit will create resistance to the airflow and must be sized accordingly. The design of a humidifier is critical and must be undertaken by someone with expertise in ventilation and humiditication.

Temperature Control

Heat from tuber respiration must be removed from the storage to keep the potato pile at the appropriate temperature. Warm air is exhausted and a small amount of cool fresh air is blended with return air to keep the potato pile cool. The quantity of cool fresh air required for temperature control is dependent upon outside temperatures. Under normal storage conditions, no auxiliary heat is required to maintain a constant pile temperature. There are circumstances however where cool outside air in excess of that required for temperature control, is brought into the storage. This situation occurs when:

- — Excess moisture must be removed from the storage. Moisture is released into the storage pile when tubers break down from disease or frost; or
- — Outside temperatures are extremely cold. Fresh air is brought into the storage to exhaust carbon dioxide and to replenish oxygen.

In these situations supplemental heat is required to warm the incoming air. A heater capacity of approximately 7.5 kilowatts per 10,000 cwt (500 tonnes) of potatoes is normally adequate. Propane or natural gas heaters must be vented to eliminate the

possibility of depleting oxygen and increasing carbon dioxide gases.

Refrigeration System

Refrigeration should be available for any potatoes stored through the late spring and summer, and can be an asset when trying to remove field heat during a hot harvest season. The basic system consists of an evaporator coil, located in the return air stream, and a compressor/ condenser unit located outdoors. The design of a refrigeration system is very complex and should only by undertaken by a person specialising in this area. The very high humidity in a potato storage requires an evaporator coil with a large surface area to prevent desiccation (drying) of the air, or even worse, frosting of the coil.

A minimum capacity of one ton of refrigeration per 2,000 cwt (0.4 kW per tonne) of potatoes is recommended. Note this capacity is only adequate for healthy, high quality potatoes. It may not maintain potatoes that are diseased or have broken dormancy.

Storage Cycle

Several distinct storage phases exist. The best management practices of each stage depend on tuber conditions, weather and the intended use of the crop.

Pre-Harvest Period

Management over the entire growing season affects the storability of potatoes. Most storage problems usually start in the field before harvest begins. Growers should aim to place a mature, disease and bruise free crop into storage. Research has demonstrated that potatoes from healthy vines are much more resistant to storage decay than potatoes from vines that have been weakened from physiological stress or from foliar diseases. Good storage design and management practices help to lessen the effects of problem tubers, but storage will never improve a poor-quality crop.

The storage should be prepared well in advance of harvest. Check all mechanical systems and clean and repair ducts. Use an accurate thermometer to check ventilation controls and check operation of the humidification system. Thoroughly clean then disinfect the storage, handling and harvesting equipment with a quaternary ammonium compound such as Ag-Services Incorporated General Storage Disinfectant, Bardak 2210 Disinfectant Sanitizer, or DMR-23 Disinfectant. Since most disinfectants are inactivated by soil and plant debris, it is essential that this material be removed by thoroughly cleaning equipment and storage with a pressure washer or steam cleaner before the disinfectant is applied. Surfaces must remain wet for at least 10 minutes for the disinfectant to destroy disease organisms.

Harvest Period

Vine killing is not employed in all types of potato production. When vine killing, desiccate or flail far enough in advance of harvest to allow potato vines to dry as much as possible, and for the skin to set. Tubers with mature skin are more resistant to mechanical damage, bruising, shrink and pathogens, which cause tuber decay in storage. Storage management is easier and the quality of potatoes coming out of storage will be better if steps are taken to reduce bruising, mechanical injury and infection from diseases at harvest.

The ideal harvest temperature is between 50 and 59° F (10° and 15°C). To avoid shatter bruises, do not harvest when the tuber pulp temperature is less than 41° F (5°C). Tubers warmer than 64-68°F (18-20°C) and under drought stress are susceptible to black spot bruising. Harvesting when tuber pulp temperature exceeds 68°F (20°C) increases the risk of leak and pink rot diseases, which can result in extensive storage decay. Don't bring severely frosted, chilled or diseased potatoes into storage.

Post-harvest Curing Period

The greatest amount of shrink occurs after harvest and before curing is complete. Harvested potatoes are skinned and there

is no barrier to moisture loss until suberin is formed over the wounds. This initial storage period promotes wound healing (suberisation) and skin set, which are critical for long-term storage quality of potatoes. The temperature, relative humidity and length of the curing period are determined by the condition of the harvested potatoes. High humidity (95%) during the curing period is critical to prevent excessive shrinkage and to promote wound healing. Mature, healthy potatoes should be cured for about two weeks at 50-60°F (10-15°C) and 95% relative humidity.

Immature potatoes with high sugars should be cured for an extended period of time at 60°F (15°C) and 95% relative humidity until fry or chip colour is acceptable. Frozen or rotting tubers should be cured at a lower temperature and relative humidity and marketed as soon as possible. The ventilation regime used during curing is determined by the cooling requirement and the need to provide fresh air (oxygen) to the tubers. Ventilate to maintain the pile at the desired suberisation temperature. You may have to ventilate more frequently if there is condensation on the surface of the tubers. Sweating or condensation on surface tubers occurs when the upper tubers are cooler than those inside the pile. A small amount of free water is usually harmless, but any excess surface moisture will encourage soft rot. Continuous ventilation is recommended when condensation is present.

Cool tubers as quickly as possible from field temperature to curing temperature. Never cool with outside air that is significantly colder than the desired storage temperature, as tubers near the air ducts will be chilled. Plenum temperature should be no lower than 3-5° F (1.5-3°C) below tuber temperature. Tubers that are harvested cold should be warmed to the desired curing temperature at a rate of about 2° F (1°C) per day, then suberised. Warm air will cause condensation on the cool tubers, making them more susceptible to soft rot infection.

Long-Term Storage

The objective of long-term storage is to maintain a consistent,

ideal environment for the duration of the storage period. Long-term storage demands more critical control than short-term storage. Recommended storage temperatures depend upon crop condition, variety and intended end use. Ventilation can be either intermittent or continuous. When storing healthy mature tubers, ventilation at intervals of 8 - 12 hours per day at full airflow rates is sufficient to maintain pile temperature. However, continuous ventilation maintains more uniform storage temperatures, particularly when outdoor temperatures are extremely low. Reducing airflow through the pile can minimise weight loss in storage. One way of accomplishing this is to operate fewer fans. Maintain a relative humidity of 95% or higher unless special storage disorders are present.

Marketing Period

Reconditioning is a storage procedure that improves chip or fry colour. Reconditioning is accomplished by increasing storage temperatures to 50-64°F (10-18°C) for two to four weeks before marketing. The higher temperatures increases tuber respiration rate, which reduces sugar levels and improves chip or fry colour. Growers are advised to consult the processor prior to reconditioning as problems can develop when tubers are exposed to reconditioning temperatures. Higher storage temperatures increase shrinkage and rot, reduce specific gravity and may break dormancy of sprout-inhibited tubers.

Potatoes may require immediate processing if serious storage decay develops. Reconditioned tubers should be processed within one month; otherwise a condition called irreversible senescence sweetening may occur. A storage facility containing several smaller (30,000 cwt or 1,360 t) bins is far more flexible with respect to reconditioning than a storage facility with only one or two 60,000 cwt (2,720t) bins. Small bins can be individually reconditioned (warmed) prior to opening and moving. Seed potatoes, marketed late in the season, should also be pre-warmed to break dormancy prior to planting and to reduce injury during handling.

CHEMICALLY IMMATURE POTATOES

Weather conditions or management practices sometimes delay crop maturity. Prompt harvest before the crop reaches chemical maturity may be necessary to avoid autumn frosts. Alternative storage management steps may be required to ensure that chemical immaturity (high sugars) of tubers does not affect processing quality out of long-term storage. Chemical immaturity can cause excessive accumulation of the reducing sugars, which results in darkening of potato products during frying. Processing tubers are usually held at storage temperatures near 59°F (15°C) with Relative Humidity near 95% for up to two weeks after harvest to promote suberisation and wound healing.

This curing or preconditioning period may need to be extended if tubers are chemically immature at harvest to prevent excessive reducing sugar accumulation in storage. During this period, sugars are either used up in respiration or are converted into starch. Depending on how much sugar is in tubers at harvest, it may take several weeks of preconditioning before reducing sugars begin to decrease and it may take several more weeks before they have dropped to a level acceptable for processing. Extended preconditioning would not be recommended for severely diseased tubers or tubers which have been damaged by frost.

Prompt marketing of such tubers would be recommended. Tubers can also be reconditioned prior to marketing, which involves gradually raising the storage temperature to lower sugars that accumulated during storage. Reconditioning just before sale may be preferred for diseased or frosted tubers since storage losses may be excessive if tubers are held at high temperatures for extended periods after harvest. Long exposure of stored tubers to high temperatures will increase the chance of weight loss due to respiration and water loss. The more physically and chemically immature the tubers at harvest, the greater the weight loss. However, the additional weight loss during preconditioning or reconditioning is justified if processing quality is improved.

Storage problems most often occur because of conditions in the field and not conditions in storage. Adverse weather, disease or improper harvesting and handling of tubers can cause problems in storage. Tubers that are rotting, frozen, chilled or diseased must be managed differently than mature, sound tubers. Good storage management will help to salvage problem tuber lots, but storage will never improve a poor quality crop.

CIPC Sprout Inhibition

Sprout inhibition is essential to maintain tuber quality for the table and processing markets. Sprouting causes tuber dehydration, physiological aging and affects the appearance of the tuber for the table market. Sprout inhibition is achieved through a combination of proper storage management and the use of a sprout inhibitor. There are two sprout inhibitors registered for use in Canada. MH60, which is applied to the crop approximately 2-3 weeks before harvest or vine kill and Chloropropham (CIPC), which is applied to potatoes after harvest. CIPC is available either as an aerosol or an emulsifiable concentrate.

CIPC aerosol may only be applied to stored potatoes by trained applicators using specialised equipment. The grower applies the emulsifiable concentrate as a direct spray onto fresh market potatoes during the grading and packaging process. CIPC aerosol, which prevents cell division, must be applied after wound healing (suberisation) and curing and before dormancy breaks. Historically, CIPC aerosol was applied with equipment originally designed for insecticide fogging.

The by-products of combustion used to generate the CIPC aerosol have negative effects on potato quality. These by-products include heat, carbon dioxide and ethylene. This problem was resolved by designing an applicator specifically for CIPC aerosol application. The new applicator technology eliminates the deleterious effects of combustion by-products and improves the effectiveness of the treatment. Both the old and new style applicators are used in Western Canada.

Storage managers should contact the custom applicator for specific recommendations before preparing the storage for CIPC application. CIPC has an affinity for water, so the humidifier should be turned off at least two days prior to application to ensure the plenum and air ducts are dry. Before applying CIPC into the potato storage, the custom applicator checks and adjusts the air system, protects the refrigeration coils and reports any factors that may affect sprout inhibition.

If storage conditions are suitable, the storage is sealed and the treatment is applied. After the treatment is complete and the fog has cleared from the storage, the applicator enters the building to perform post treatment cleaning of fans, fan guards, plenum, and other surfaces in contact with the chemical. Sealing materials are removed and the storage controls are reset to pre-treatment conditions, allowing the grower to resume normal operations. Warning: Only trained personnel wearing protective clothing should enter the storage during treatment.

Occasionally, CIPC treated potatoes sprout prematurely. Poor performance of the sprout inhibitor could be caused by:

— Temperature fluctuations and "hot spots". These increase the rate of tuber respiration resulting in poor sprout inhibition. Temperature fluctuations and "hot spots" are caused by an improperly designed or malfunctioning ventilation system or excessive dirt and tuber rot in the bulk pile.

— Physiological stress. Field-stressed potatoes may respond differently to CIPC application than potatoes grown under normal conditions. Studies have shown that potatoes grown with deficient nitrogen will sprout earlier than adequately fertilised potatoes. Other field stresses may also reduce the effectiveness of the sprout inhibitor.

— A low CIPC concentration on the sprout can cause internal sprouting where the sprout grows inward into the tuber or outward into an adjacent tuber. This defect occurs mainly in long-term storage. Late application,

pile settling or the presence of excessive dirt and debris may interfere with the application, resulting in a low concentration of CIPC on selected tubers.

— Improper application.

— Poor storage management including fluctuating temperatures, high levels of respiration gases, low humidity, etc. shorten the dormancy period and reduce the effectiveness of sprout inhibition.

Contamination of seed lots by CIPC can occur one or more years after a storage treatment. Do not store seed potatoes in a structure that was recently treated with CIPC or in a structure adjacent to a building where CIPC will be applied. There are no effective methods to rapidly decontaminate a storage structure after CIPC application. Seed can only be safely stored in a structure one year after the fans, ducts, and plenums are thoroughly cleaned of all CIPC residues and warm air has been circulated through the storage during the summer period.

Food Safety

In the process of growing, harvesting, storing and shipping, potatoes are exposed to many conditions that may result in food-safety problems. There are three general types of contaminants that can either reduce quality or actually make potatoes unsafe for consumers. These are:

— Organisms, such as bacteria, but also larger animals like mice.

— Objects, (foreign materials) such as glass, metal, wood or plastic

— Chemicals, such as disinfectants, cleaners, fuels or pesticides.

Contaminated potatoes can cause illness or injury to a consumer. The presence of physical contaminants can cause losses and equipment failure in processing and packing plants. Awareness, documentation of all practices and a safe-food attitude are important in ensuring food safety on the farm.

The following are methods for preventing contamination by microorganisms and other pests:

— Potato storages must be sanitised before potatoes are stored. Walls, floors, plenums and ducts should first be washed with high-pressure water and then a general storage disinfectant.

— Storage cleaning records, which include date and materials used, should be kept. Name of the person doing the cleaning should be recorded.

— Where diseases are known to be present in stored potatoes, trays containing disinfectant (foot baths) should be placed at storage entrances to reduce the risk of spreading diseases to other storages.

— Animals such as mice, gophers, farm pets, and birds must be prevented from entering storages so their feces and carcasses do not contaminate potatoes.

— All work and storage areas should have adequate toilet facilities including hand-washing facilities that include liquid soap dispensers, warm water and disposable paper towels for drying. Staff must wash hands after using facilities.

— Producers who have both livestock and potato production facilities should have separate footwear for each facility.

— All wash water should be clean and free of microbiological and chemical contaminants.

— Cull piles are a potential source of potato diseases and contamination of stored potatoes. Cull piles should be disposed of away from potato facilities as soon as possible.

The following practices prevent contamination with foreign materials:

— Loose metal objects such as nuts, bolts and nails must be removed from storages prior to storing potatoes. Small metal objects can embed in individual tubers and pass undetected through grading facilities.

— If a grower suspects there could be foreign objects in potatoes, the packer or processor should be alerted. One small piece of metal can cause a costly plant shut down. It also becomes a safety hazard for consumers if undetected.

— Glass cannot be detected with scanning devices in processing plants and is therefore a major concern.

— Standard light bulbs should not be used anywhere in potato storages where glass could fall onto potatoes. Shatter-proof light bulbs should be used over grading lines and in storages or protective shields must cover bulbs.

— Broken glass must be thoroughly cleaned up in a storage complex.

— Plastic film or small plastic objects are potentially as dangerous as metal and cannot be detected with metal detectors common in processing plants.

The following practices prevent contamination with chemicals:

— Chemicals should be stored in well-ventilated areas away from storages. Pest control chemicals must never be housed in a potato storage or receiving area.

— Storage should not be used as a shop or garage at any time.

— If vehicles are temporarily stored in potato storage and handling areas, all traces of grease and oils must be removed prior to storing or handling potatoes.

— All spilled chemicals including oils, solvents, lubricants and cleaners must be thoroughly cleaned.

— If diesel or gasoline fumes are noticeable, the area must be thoroughly aired prior to storing potatoes.

— Pre-harvest or pre-marketing intervals must be adhered to when applying pest control products and sprout inhibitors.

— Use only registered pest control chemicals on potatoes in the field or in storage.

- — Sprayers or chemical applicators must be calibrated to ensure rates are accurate. Date of calibration should be recorded.
- — Thorough and accurate records should be kept for all fertilisers and pest control products applied to potatoes. Names of applicators should be recorded. Record books specifically for chemical applications are available from dealers and processors. Records of products applied should include information such as:
  - — Fertiliser or chemical supplier
  - — Type of fertiliser and rate applied
  - — Pest-control chemical names and rates
  - — Land locations where products are applied
  - — Dates of application
  - — Name and initials of the applicator

General practices to ensure a safe supply of potatoes:

- — Documentation is key for a successful programme and will help a producer in the event of a problem.
- — All areas and facilities where potatoes are handled or stored must be clean and sanitary.
- — Develop and maintain a safe food attitude on the farm.
- — Learn how to identify and control potential hazards on the farm.
- — Be aware of potential risks and constantly be on the lookout for hazards to food safety.

Develop a working environment where staff is encouraged to make suggestions or point out potential risks to the safety of the potatoes grown and stored.

Preparing Ware Potatoes for Storage

All potatoes showing greening, any decay or damage should be rejected for storage. Immature tubers and those showing minor damage or wetting by rain may be put aside for immediate consumption. As with other root crops potatoes to be stored need to be cured to repair any skin injuries and to

promote the formation of a stronger epidermis to reduces water loss. The optimum conditions for curing potatoes are:

Temperature: 15° - 20°C

Relative humidity: 85% - 90%

Duration: 5 - 10 days

If a sprout inhibitor is to be used, it should be applied after the curing operation has been completed otherwise curing will not take place. Sprout inhibitor should neither be applied to, nor come into contact with, potatoes to be used as seed.

Low-Cost Storage System

Simple low-cost storage of ware potatoes is possible in a wide range of structures and conditions provided that the basic principles of store design, operation and management are followed. In the highland tropics ambient temperatures of 10°C to 15°C at night and 15°C to 18°C during the day make it possible to use very simple methods of potato storage. In the lowlands of Sub-Sahelian Africa where potato growing has only recently been introduced there has been no general development and significant adoption of satisfactory storage systems, although work in this field is still on-going.

Clamp System

The clamp is a simple and inexpensive storage method particularly applicable to the more temperate regions of Europe and Latin America. The details of design vary considerably but all follow similar lines. CIP recommends that tubers are heaped on a bed of straw 1m to 1.5m wide with a ventilating duct on the floor in the centre of the heap and covered with a layer of about 20 cm of compacted straw or 30 cm if not compacted. The clamp must be in a well-drained location. Temperatures inside a ventilated clamp will be approximately those of the ambient temperatures. The clamp system can be modified to meet local requirements provided nothing is done that will lead to an unwanted increase in temperature. To avoid too high a temperature in warm climates the clamp should be shallow, but the length can be

increased if necessary to maintain capacity. Where rodents or other mammals are a nuisances a wire net can be dug at least 15 cm deep into the ground and covered over the clamp.

Naturally Ventilated Ware Potato Stores

There must be sufficient movement of cool air through a heap of potatoes to remove respiratory heat and keep the potatoes as cool as ambient temperatures will allow. The design of any naturally ventilated store is based on the principle of natural air convection. This type of store has been the subject of intensive extension programmes by FAO for farmers in the highlands areas of Cameroon and by CIP in Kenya. The critical factor in maintaining the store temperature as near ambient as possible is the size of the heap. Generally, the maximum height of the pile of potatoes should not exceed 1.2m. If, for any reason store temperatures are expected to be higher than 25°C, it is better to keep the potatoes in small lots, in small crates or in shallow layers.

Cameroon

The Naturally Ventilated Store introduced by the FAO project was a mudblock structure, 1.2m long, 1.2m wide and 1.0m high to the eaves with a thatched roof and a capacity of about 700kg The capacity could be increased by making the store longer. The lower part of the store was a ventilating chamber or plenum made over a pit about 35cm deep into which was built a stone foundation wall 15 cm deep to support a mud block wall on the periphery of the pit and rising to ground level. Wooden sticks (saplings or sawn timber) were spaced two or three centimetres apart and laid over the stone wall to make a ventilating floor.

A mudblock wall was continued upwards for a further one metre to the eaves. On the windward side of the structure a ventilation opening was fitted with a shutter to control ventilation. The storage section was from the false floor up to eaves at one metre high and was provided with closeable openings for loading and emptying. The roof was a structure of raffia covered with grass. Store management was simple.

After curing the potatoes are treated with a sprout inhibiter. During the storage period the shutter on the ventilation opening is left open at night when ambient temperatures are lowest (13°-18°C in Cameroon) and closed during the day when temperatures are higher and relative humidities lower.

A series of trials, carried out both at research station and farmer level, compared the FAO store with prevailing storage methods. The results clearly indicated that the cool night air ventilation and reduced day time ventilation lead to a significant improvement in the condition of potatoes after storage over the existing traditional methods. Good quality potatoes could be stored for four months with losses of about 10%.

The advantage of the improved store to farmers was that instead of being forced to sell all their crop within a maximum period of two months when market prices are at their lowest level, this new store allowed them to extend or delay the marketing of their potatoes until prices were more favourable.

The construction of the store is limited for the most part to locally available materials, mud blocks, bamboo or wood poles, wooden shutters and few locally available carpentry items. The cost varies from 5,000 francs CFA, when all materials, excluding carpentry items, were available on the farm, to 18,000 francs CFA when the farmer pays for all the materials and labour.

The cost/benefit ratio was estimated to be 2.5, on the basis of information provided by farmers who had adopted this technology. Thus a capital investment of 18,000 CFA could be recovered after two storage periods.

Ventilated warehouses in cameroon

The principle of naturally convected air ventilation was applied to the storage of potatoes in warehouses of 7-20 tonne capacity. The structures were built of mudblock walls with a galvanised corrugated sheet steel roof. Controllable ventilation louvres, each 70cm x 30cm, were built into the two long walls, those at the windward side were 50 cm above the

ground, those on the opposite wall were built at about 30 cm below the eaves level.

There was at least one pair of louvres (inlet and outlet) for every three linear metres of wall. A ceiling of bamboo or raffia was built under the roof on which was placed a 30 cm thick layer of packed straw to insulate the building from the heat radiated from the metal roof.

Mudblocks were preferred to cement blocks because their insulation was greater. The ventilation louvres were opened only at night. The average internal store temperature was 14.5°C with a mean minimum outside air temperature (at night) of 10°C and a mean maximum outside air temperature of 30°C.

More than 50 stores of this type were operated by women organised into pre-cooperatives for the marketing of potatoes. The potatoes were placed inside small crates (50 x 30 x 30 cm) which were stacked in rows inside the store, leaving inspection alleys of 80 to 100 cm, wide. Each crate was marked with the name of its owner but marketing was organised as a group.

Seed Potato Storage

The effect of light on reducing sprout elongation and the physiological ageing process has long been known. CIP and others have demonstrated how a store using natural diffused light can be used for storing seed potatoes by farmers in tropical countries.

Only a simple shed protecting the potatoes against rain and direct sunlight, and equipped with shelves on which the potatoes are spread in shallow layers is needed. After few weeks of storage, seed tubers turn greenish and develop short, sturdy and green sprouts. Protection against rodents attacks may be necessary.

Farmers in tropical regions generally consider sweet potato to be difficult to store, especially in areas where Cylas formicarius, the sweet potato weevil or a related species, Cylas puncticollis are major threats. Because of this threat in many

cases the crop is left in the ground and harvested only as needed. Sweet potato can be stored safely for 3 to 4 months in the tropics provided that the storage practices previously outlined for potatoes are followed. These include:

- — selecting for storage only the fully mature and undamaged roots;
- — curing the roots at 30° to 32°C and at 85% to 90% relative humidity for four to seven days;
- — handling the roots carefully and limiting their handling to a strict minimum;
- — keeping the storage temperature at the lowest level possible (an optimum of 13°C);
- — keeping the relative humidity as close as possible to 85%-90%.

PRINCIPLES OF STORAGE FOR ROOTS AND TUBERS

Root and tuber crops need to be handled gently to minimise bruising and breaking of the skin because of its relatively soft texture compared, for example, to cereal grains. The effect of mechanical injury resulting in external and/or internal bruising and tissue discolour-ation is often underestimated. Severely damaged tubers should not be stored for three reasons;

- — because of lower quality
- — because of the increased risk of subsequent pathogenic losses.
- — because of the risk of introducing disease organisms into sound produce.

Most mechanical damage occurs as a result of careless handling at harvest and during transport to and within a store since, generally in the tropics, food handling procedures are poorly developed and fresh produce is all too frequently treated as an inert object. Careful harvesting and proper handling of roots and tubers is, therefore, an important step towards successful storage. Crops are most likely to be injured at harvest by the digging tools, which may be wooden sticks,

machetes, hoes or forks. Therefore, immediately after harvest, the crop must undergo the operation of curing. Roots and tubers have the ability to heal their skin wounds when held at relatively high temperatures and humidities for few days after harvest (Table 1) whilst at the same time there is a general strengthening of the skin.

The term "curing" refers to the operation of self-healing of wounds, cuts and bruises. There are two steps involved in the curing process:

— Cells suberisation. The production of suberin and its deposition in cell walls;
— The formation of cork cambium. The production of cork tissue in the bruised area. The new cork tissue seals the cut or bruised areas and helps prevent the entrance of decaying organism while also reducing water loss.

Table 1. Conditions required for curing root and tuber crops

Crop	Temperature °C	Relative humidity (%)	Duration (days)
Potato	15 - 20	85 - 90	5 - 10
Sweet Potato	30 - 32	85 - 90	4 - 7
Yam	32 - 40	85 - 95	4 - 7
Cassava	30 - 40	90 - 100	4 - 4

Although wound healing is important in preventing the invasion by pathogens, it also important in limiting the rate of respiration and water loss.

Table 2. Percentage weight loss of cured and uncured root crops (Booth 1972)

Crop	Duration of Storage (days)	Percent Weight Loss	
		Cured	Uncured
Yams	150	10	24
Sweet Potatoes	113	17	42
Potatoes	210	5.0	5.4

Since water loss in badly suberised and damaged tubers may be very high, it is important that the curing operation is carried out as soon as possible after harvest. This will encourage as complete healing as possible although the extent of healing will depend on the on the type of damage; the older the tuber, the less well a wound will heal. Four factors affect the healing of wounds;

— temperature of the commodity

— oxygen and carbon dioxide concentration within the commodity

— humidity within the commodity

— the use of sprout inhibitors

It is perhaps unfortunate that the curing operation when carried out properly involve conditions of high temperature and high humidity within the commodity, which also favour the rapid growth of microorganisms and commodity deterioration.

Therefore it needs to be carried out very carefully and quickly and it follows that after curing further handling of the produce should be reduced to a minimum so as to avoid further injury that may result in further losses. The length of time for proper curing cannot be precisely defined as it depends on many factors, such as condition of the crop at harvest, type of wound, season, storage temperature and relative humidity.

Types of Packaging Material

The ideal in packaging is to protect the produce from damage during handling, transport and storage and to provide containers of uniform size that are conveniently stacked and handled, easily accounted for in quantity and, where appropriate, in weight.

In many developing countries traditional baskets, and various types of trays or buckets are used for transporting produce to the house or to village markets. These are usually of low cost, made from readily available material and serve

the purpose for transport over short distances. But, they have many disadvantages in large loads carried over long distances:

— they are difficult to clean when contaminated with decay organisms;

— they often lack rigidity and distort when stacked thus applying severe local pressure to their contents

— they are frequently very variable in shape and therefore are difficult to load, especially for long journeys.

— being of local manufacture they are often rather crude and may have sharp edges or splinters causing cuts and punctures to the commodity.

Many authorities have observed that produce being transported and marketed in commercial quantities needs better packaging in appropriately sized units if losses are to be minimised and to achieve economical use of transport.

The shape of packages is significant because of need to load for maximum capacity and stability. Round baskets, whether cylindrical or tapered, hold considerably less produce than boxes occupying the same cubic space; a cylindrical basket contains only 78.5% by volume compared with a rectangular box occupying the same space on a vehicle.

However, packaging can be a major item of expense in produce marketing, especially in developing countries where packaging industries are not well developed. The selection of suitable containers for commercial scale marketing requires very careful consideration.

Among the various types of packaging material that are available, natural and synthetic fibre sacks and bags as well as moulded plastic boxes seem to be more suitable and have greater promise for packaging roots and tubers and for their transport to distant markets.

Natural and Synthetic Fibre Sacks

Bags up to 5 kg capacity and sacks from 25 to over 50 kg can be made from natural fibres, jute or sisal, or from synthetic

fibres of polypropylene or polyethylene. These sacks are ubiquitous and have been in continuous use for packaging and transporting root and tuber crops to markets for a considerable time.

Their main problem in many developing countries is that often they are too large and when filled they are too heavy for careful handling and are often dropped or thrown down resulting in severe bruising and damage to the contents. Finely woven sacks, particularly of synthetic material impair ventilation when stacked.

Moulded Plastic Boxes

Moulded high-density polythene boxes, allegedly reusable, are widely used for transporting perishable produce in many countries. They can be made to almost any specifications, are strong, rigid, smooth, easily cleaned and can be made to stack when full of produce and to nest when empty in order to conserve space. Their disadvantages are:

- they can be produced economically only in large numbers but are still costly;
- they have to be imported into most developing countries, adding to the cost and pressures on foreign exchange;
- there has to be a good organisation and sound reliable cooperation between the market traders, transport services and producers if there is to be a effective container return service.
- the containers, particularly the cheaper varieties, deteriorate rapidly when exposed to sunlight, unless treated with an ultraviolet inhibitor.

Before deciding on what packaging to use, the producer or trader has to ensure that the cost of packaging does not exceed its benefits. A decision should involve consultation between the transport contractors and market traders. The factors to be considered include;

- the type of produce;

— the level of losses occurring during marketing;
— the comparative cost of the present and improved packaging;
— the regularity of supply of the packaging material;
— the acceptance of the packaging method to the market.

Temperature Control

Temperature has a great influence on many factors that cause loss during storage. It is the single most important factor affecting the rate of respiration, it also influences the rate of sprout growth, the development of rotting microorganisms and insect infestation.

At 10°C, the rates of sprout development, rotting and respiration are shown to be moderate but at 4°C, sprouting is stopped, while rotting and respiration continue but at very low levels. With the exception of highland areas, low temperature storage in the tropics within the range 10° to 15°C can only be envisaged by using a refrigeration process of some kind.

At subsistence or small farmer level this is generally not practical because of cost implications and the technical support needed to sustain conventional refrigeration technology. Only in very dry areas is simple evaporative cooling at all successful but even this simple technology needs a prime mover to be operating almost continuously.

Therefore, successful storage of roots and tubers in any sort of structure depends very much on natural ventilation to remove respiration heat, to remove carbon dioxide, which in concentration can lead to the breakdown of dormancy, and to keep the temperature of the crop as low as possible. Ventilation should be with the coolest possible air, night time ventilation is not only the coolest but has the highest relative humidity, so that water loss through transpiration is also held to a minimum.

Gamma Irradiation

Gamma irradiation is known to induce lesions on nucleic acids and cellular proteins thus preventing them from multiplying.

When applied at doses of about 7.5 head, gamma irradiation has been reported to inhibit sprouting of yams, potatoes and sweet potatoes However, this technique has not yet been applied on a commercial scale in the tropics and is unlikely to be of practical value to farmers because of the cost of the high technology involved.

Table 3. Monthly cumulative percentage loss in weight of irradiated yam tubers, the radiation.

Storage Duration Months	12.5 (krad)	10.0 (krad)	7.5 (krad)	5.0 (krad)	2.5 (krad)	Control
1	1.4	2.5	2.8	2.5	4.6	3.5
2	3.8	5.8	6.2	6.5	8.5	9.2
3	5.8	8.0	8.5	9.0	15.9	19.5
4	8.8	11.2	12.0	12.6	23.7	25.7
5	10.7	14.6	15.9	16.4	29.8	35.6
6	12.8	16.5	18.2	20.2	35.5	41.2
7	15.9	18.6	21.9	25.1	41.8	49.9
8	18.6	20.2	24.8	28.0	45.7	55.9

Furthermore it is a treatment that is not widely acceptable to consumers or a permitted treatment for food commodities in some developed countries.

Control of the Spread of Diseases

Efforts to control the spread of diseases should aim to be preventive rather than curative. Simple and low cost preventive measures which to help control the incidence of post-harvest diseases include:

- gentle handling to minimise the risk of injury to tubers during harvesting, transport and storage;

- adequate cultural practices and especially using disease free planting material;
- good phytosanitary practices, including regular inspection of fields and premises, proper disposal of diseased tubers and plant debris, the cleaning and sterilising of implements, boxes, buildings, etc.
- pre-harvest crop application of chemicals to control the diseases in the growing crop;
- curing of the crop before storage;
- only storing produce that has been dried before putting into store, avoiding produce in store getting wet and storing at the optimum temperature.

Some diseases can be prevented or controlled by the direct application of chemicals to the produce by dipping the produce, applying sprays or dusts. To be effective chemical treatment requires the application of the appropriate compound, at the recommended dosage, by the most appropriate method and at the most suitable time.

It is important to have a thorough knowledge of the pathogens being treated in order that the correct treatment can be selected and properly applied.

Thiabendazole and Benomyl are presently the most commonly used fungicides for post-harvest treatment of pathogenic diseases of roots and tubers. The best results are obtained when the chemicals are applied not later than three days after harvest and not after the pathogen is well established.

One drawback to these particular chemicals is that they are not always readily available to farmers or they are too expensive. There is a very real risk, which is frequently borne out in practice, that farmers who, for any reason, cannot obtain these chemicals will use what ever is at hand without regard to the danger of applying chemicals which are suitable for growing plants but which are dangerously toxic when applied directly to stored food produce, whether or not the produce will be treated, as a minimum being washed in potable water, before being consumed.

Control of Damage Caused by Insects

Insect pests can be the cause of serious losses in stored roots and tubers, yams and sweet potatoes in particular. Surveys carried out in 1981, 1983 and 1984 in Côte d'Ivoire showed increasing levels of infestation of stored yams over a period of four months storage, with 63% of stored tubers being infested by moths and weight losses of 25% attributed to insects.

Good hygiene is of paramount importance in insect control including, particularly, the destruction by burning of infested tubers and rubbish that can act as host to a variety of insect pests and cleaning and disinfection of the store structure.

In many areas it may still be necessary to use some form of chemical control especially if storage is extended over several months. Various methods of control of the potato tuber moth in potato stores have been tested in many countries.

In stores in Bangladesh dried and crushed Lantana camara, as well as the insecticide Decamethrin (Decis), has been reported to be effective. In Kenya moth damage was reduced significantly by the repellent weeds Lantana camara, Minthostachys sp. and Eucalyptus sp. Deltamethrin used as a spray of 2.5g active ingredient per 100 litres of water, has been reported to be effective in controlling moths (Tineidae sp) on stored yams.

Insecticides may be applied as dusts on the planting material, on the soil during the tuber-forming period, or as sprays applied to the growing crop. The same observations on the misuse and application of unsuitable chemicals apply equally to use insecticides as to fungicides. Allegedly safer and more practical alternative control methods are continually being developed which are particularly suitable for subsistence agriculture.

Some level of control can be achieved through the use of established insect repellents, such as "lantana" and through crop rotation and forms of shifting cultivation which reduce

the likelihood of a serious build-up of soilborne pests. Another promising control is the use of cultivars that have a marked genetic resistance to insects and new cultivars are being bred with resistance characteristics.

Control of Nematodes

Nematodes are well known to cause serious losses of roots and tubers, of both quantity and quality. There are three possible methods of control:

— Application of chemicals to soil and plants. This treatment has not proved to be economical;

— Treatment of propagative material prior to planting by immersion in nematicides or hot water;

— Cultural practices. The most common and reliable method. It involves growing alternative crops for several years which are not suitable hosts for the specific nematode pest. It is important that the alternative crops really are unsuitable hosts for the nematode and that there are no ground keepers and weeds left growing that will permit a colony to survive.

Of course, the ideal solution to the problem of nematode infestation is to grow nematode-free seed material in soils from which nematodes have been eliminated.

References

Brook, R. C., R. J. Fick, and T. D. Forbush, *Potato Storage Design and Managemen*, East Lansing, MI.: Michigan State University Agricultural Engineering Information Series, AEIS#627, 1995.

Christ, B. J., *Potato Diseases in Pennsylvania*, Pennsylvania College of Agricultural Sciences, University Park, PA. 1998.

Greaser, G. L. and J. K. Harper, *Agricultural Alternatives: Enterprise Budget Analysis*, University Park, PA: Penn StateCooperative Extension, 1994.

Lamont, W. J., *et al.*, *Agricultural Alternatives: Irrigation for Fruit and VegetableProduction*, University Park, PA: Penn State CooperativeExtension, 2001.

Rowe, Randall, editor, *Potato Health Management*, St. Paul,MN: APS Press, 1993.

Sieczka, J. B. and R. E. Thornton, *Commercial PotatoProduction in North America*, Orono, ME: The PotatoAssociation of America Handbook, 1992.

Weisz, R., Z. Smilowitz, M. Saunders, and B. Christ, *Integrated Pest Management for Potatoes*, University Park,PA: Penn State Cooperative Extension, 1995.

13

Factors Affecting Storability of Potatoes

The skin of a mature root or tuber is normally an effective barrier against most potentially invading bacteria and fungi causing rotting of the tissues. Any rupture of this barrier caused by damage or injury to the skin will provide an entry point for infection and will also stimulate physiological deterioration and dehydration. There are different degrees of mechanical damage, from small bruises to deep cuts and they may be sustained at any stage, from pre-harvest operations, through harvesting and subsequent handling operations when the product is graded, packed and transported for market or, simply, even carried to the farmer's house. Serious mechanical injury, which may result in the product being rejected during grading, is a direct loss. Damage to the tuber skin that is not immediately obvious can lead to physiological deterioration and allow the entry of pathogens.

Respiration Process

Roots and tubers are living organisms and as such, they respire. The respiration process results in the oxidation of the starch contained in the cells of the tuber, which converts it into water, carbon dioxide and heat energy. During this transformation of the starch the dry matter of the tuber is reduced. The respiration process can be approximately represented by the oxidation of glucose:

$$C_6H_{12}O_6 + 6O_2 \rightarrow 6CO_2 + 6H_2O + \text{energy}$$

For respiration to occur freely a supply of oxygen is needed and the resulting carbon dioxide and heat have to be removed from the products' environment. A limited supply of oxygen and inadequate removal of carbon dioxide may lead to effective asphyxiation and the death of product tissue. Ideally the complete combustion of 1g of glucose yields 1.47g CO_2 + 16 kJ of energy. In practice about 5.1 kJ (32%) of this energy is used as metabolic energy and 10.9 kJ (68%) is released as heat. The rate of respiration is assessed either by measuring the uptake of oxygen or the quantity of carbon dioxide released and is expressed in milligrams of CO_2 per kilogram of tuber per hour.

Respiration Rate

During its physiological development through growth, harvesting, storage and subsequent planting as seed, the living tuber passes through different stages during which the respiration rate varies with;

- — the physiological age of the tuber,
- — whether it is sprouting or in its dormant phase,
- — whether or not it has been damaged and is healing its wounds,
- — the storage conditions, mainly temperature.

Generally, the rate of respiration is relatively high at harvest, followed shortly by a decrease, especially during storage, then followed by an increase once sprouting begins. For example, the dry matter losses for potatoes stored at 10°C are approximately 1% to 2% during the first month after harvest, 0.8% per month during storage but rising to 1% to 5% per month when sprouting is well advanced.

Physiological Age

During dormancy, the endogenous metabolic rate of tubers is at its minimum and the dry matter losses are correspondingly reduced. For example, it has been shown that immediately after harvest yam tubers respire at a rate of 15mL CO_2/kg fresh weight/hour at 25°C.

Table 1. Loss of weight in Yam attributable to respiration

Age of Yam tubers (% per day)	Total weight loss due to Respiration		Percent of total weight loss	
	25°C	35°C	25°C	35°C
After harvesting	0.22	0.36	27	30
During dormancy	0.15	0.28	7	10
During sprouting	0.21	0.34	35	20

The respiration rate will later drop as low as 3 ml CO_2/kg/hr and remain at that level until sprouting starts. During sprouting the respiration rate increases dramatically to over 30mL CO_2/kg/hr.

Skin Permeability

The permeability of the skin of the tuber is a function of its maturity and is a very significant factor in the rate of respiration. The periderm of freshly harvested immature tubers is most permeable and thus permits greater levels of respiration than similarly harvested mature tubers. Immature potatoes are reported to respire at a rate of about 17mL O_2/kg/h immediately after harvest, compared to a rate of 5ml O_2/kg/h when physiological mature.

Skin Damage

Damage and wound healing greatly influence respiration. It has been found that simply cutting through a potato will double its respiration rate and dropping the potato from a height of about 1 metre will so damage a potato skin as to increase respiration by 30% to 50%. For sweet potatoes the rate of respiration of a damaged tuber doubles in response to a wound after a delay of about 20 hours. This effect is known as wound respiration response.

Storage Temperature on Respiration

Temperature is the single most important factor affecting the rate of respiration. For biological material this can be

expressed as; $Q_{10} = 2$, i.e. the rate of respiration is doubled for every 10°C increase in temperature over the range 5°C to 25°C. Significant variations to this general rule will occur through interactions with other factors affecting the rate of respiration, as discussed in the previous paragraphs.

For potatoes, the rate of respiration is at a minimum around 5°C. Below this temperature the respiration rate tends to increase. For yams, it is known that low temperatures reduce the rate of metabolic activity of the tubers (Table 1), but temperatures in the range 10° to 12°C cause damage through chilling which, because of a breakdown of internal tissues, increases water loss and increases susceptibility to decay. Very low temperatures, below 5°C, similarly affect sweet potato and cassava. The symptoms of chilling injury are not always obvious when the tubers are still in cold storage, they become noticeable as soon as the tubers are restored to ambient temperatures.

Transpiration of Water from Tubers

Transpiration is water loss through the skin pores of the tuber and is, effectively, evaporation. Because roots and tubers are characterised by having high moisture content, even in the ambient conditions prevailing in the humid tropics, they will continually lose water to the surrounding air. This loss of water can be significant in several ways.

Whilst the original food value may not be affected a large water loss will adversely affect the quality of the produce, for example a loss greater than 10% will result in a bigger peeling loss because of the shrivelled texture of the skin and the culinary quality may also be affected. A weight loss becomes an economic loss when the produce is marketed on a weight basis as well as being less attractive to potential customers.

It is very difficult to give precise data on transpiration weight loss, even under precisely defined conditions because of the wide range of factors by which it is influenced. These include:

— temperature,
— relative humidity,
— the rate of air movement surrounding the tuber,
— most significantly, the permeability of the skin and how this may have been affected.

Temperature and Relative Humidity of the Air

There are interstitial spaces within all plant tissues that allow water vapour and air to move throughout the plant. The water vapour within these spaces exerts a water vapour pressure, which pressure is a function of the amount of free water contained within the tissues and its temperature. The rate at which water is lost from fresh tubers depends on the difference between the water vapour pressure within the tubers and the water vapour pressure of the surrounding air, with moisture passing from the higher pressure to the lower.

If there is a considerable difference between the temperature of the produce and the surrounding air, temperature becomes the dominating influence on water vapour pressures, whereas when both are at similar temperatures it is the amount of water vapour that has the most significant effect. It therefore follows that to minimise water loss from high moisture content produce, the produce should be kept in atmospheres that have a comparable water vapour pressures. In practical terms this means in a cool moist atmosphere.

Air Movement and Loss of Water

The greater the velocity of air moving over fresh produce the faster is water lost though transpiration. Air movement through produce is essential to remove the heat and CO_2 produced by the respiration of the produce but the rate of air movement needs to be kept as low as practical to prevent excessive loss of moisture.

Permeability of Tuber Skin

The skin of tuber crops will allow water vapour to diffuse

through it during respiration and transpiration. The factors discussed earlier which affect the rate of respiration also affects the rate of transpiration through the skin. At 10°C, undamaged mature potato tubers are reported to lose water at a rate of 62 to 109 ug/cm^2/hr/kPa, while immature tubers harvested 24 hours previously lost water at a rate of 1-6mg/cm^2/hr/kPa, a factor 15-100 times greater.

Dormancy and Sprouting

Yam, cocoyam, potato and sweet potato tubers propagate vegetatively. To counter what is often an unfavourable climate at the end of their growth period, they go into a dormant phase. The beginning of this period is considered to be the point of the physiological maturity of the tubers, also called "wilting point". The dormancy period can be defined as the period of reduced endogenous metabolic activity during which the tuber shows no intrinsic or bud growth, although it retains the potetial for future growth.

Dormancy is both a species and a varietal characteristic. It is also affected by other factors, temperature is the most important but others, including moisture, oxygen and CO_2 content of the storage atmosphere, the extent of wounding and any disease of the tuber, real or putative, although normally of lesser importance may, occasionally, have an over-riding effect. The cassava root, as opposed to other roots and tubers, is a plant of perennation and not propagation. It has no dormancy and it senesces soon after harvesting.

Passam suggested that differences in the dormancy of yam species are the result of the ecological environments in which the different species have evolved. For example, varieties of yam native to regions with marked arid seasons have a longer period of dormancy than those that are native to regions with shorter dry seasons. D. cayenensis, which originates from the West African forest zone where the dry season is very short, shows almost continuous vegetative growth.

In contrast, D. alata and D. rotundata, originating respectively in Asia and Africa, appear to be adapted to

climates where there is a longer dry season during which the plant survives as a resting tuber. These inherent differences in dormancy are responsible to a large extent for variations in the ability of different species to store well.

Temperature Effect on Dormancy

Lower storage temperatures are widely practiced as a technique for reducing the metabolic activity of roots and tubers and prolonging their dormancy. Temperatures of 16° to 17°C have been used to prolong the storage period for D. alata tubers for up to four months, provided the tubers were properly cured prior to storage in order to control infection by wound pathogens.

Table 2. Dormancy periods of the major edible yams

Yam species	Locality	Length of dormancy (weeks)
D. alata	Caribbean	14 -16
	West Africa	14 -18
D. rotundata	West Africa	12 - 14
D. cayenensis	West Africa	4 - 8
D. esculenta	West Africa	12 - 18
	Caribbean	4 - 8
D. trifida	Caribbean	4

For potatoes, sprout growth is practically negligible at 4°C and below and increases with increasing temperature. However, avoidance of sprouting by low temperatures leads to sweetening of the tuber, which is considered to lower the value of the stored crop.

End of Dormancy

While roots and tubers remain dormant they can be stored satisfactorily. As soon as dormancy is broken and sprouting begins, the rate of dry matter loss increases dramatically since the formation of sprouts requires energy, which is drawn from the tubers' carbohydrate reserves. The rate of water loss also increases and if this becomes excessive the tubers dry out

allowing pathogens to penetrate the tuber, potentially causing severe damage if not total loss, making continued storage quite impracticable.

Effect of Pathological Factors

All living organisms are subject to invasion by microorganisms, fungi, bacteria and viruses, which constitute the most serious cause of direct post-harvest loss in tropical root crops. These disease organisms are widely distributed in the air and soil and on dead and decaying plant material.

Table 3 Cumulative percentage of sprouting of yams tuber stored at different temperatures

Storage period (Months)	15°C	20°C	25°C	Yam Storage Barn (ambient)
0	0	0	0	0
1	0	14	0	0
2	0	40	28	30
3	0	54	94	88
4	0	54	100	100
5	0	56	100	100

The extent of the occurrence and the magnitude of losses due to pathogenic microorganisms are very variable. The time of infection varies with the crop and with different diseases, it can occur in the field before harvest or at any time afterwards. Since many post-harvest pathogens are introduced through wounds, one of the major factors governing the incidence and magnitude of such losses is the physical condition of the produce. The cork layer surrounding the roots and tubers is intended to serve as a barrier against bacterial and fungal attack.

Infection Sources

The infection may start:

— Before harvest, through natural pores in the above and below ground parts of plants, which allow the

movement of air, carbon dioxide and water vapour into and out of the plant.

- Due to injuries caused after harvest by careless handling, by insect or other animal damage.
- By direct penetration of the intact skin of the plant; the time of infection varies with the kind of crop, its maturity and the type of disease organism; it can occur in the field before harvest or at any time afterwards.

Pre-harvest field infection does not necessarily become apparent until after harvest but can occur at any time between the field and the final consumer. The infecting organism may be distributed by infected seed or other planting material, from crop residues or rejected produce left decaying in the field or field boxes or packhouses, from contaminated water used to wash produce and by contaminating healthy produce from diseased produce in the same package. Many diseases can survive by using weed plants or other crops as alternative hosts.

Pathogenic Loss

Losses reducing the quantity of sound produce are, generally, the more serious losses but are often underestimated because they are not easily recognised or evaluated. They are often caused by infection of the produce in the field before harvest either by a primary infection or a secondary infection following an initial infection by one of a few specific pathogens, normally through a wound. The initial infections cause a breakdown of the host tissues and once these primary pathogens are established, they are followed by an invasion by a broad spectrum of secondary pathogens.

Losses affecting the quality of the produce and which occur when the disease affects only the surface of the produce do not necessarily affect the intrinsic quality or quantity of the commodity but makes the crop less attractive to the consumer or buyer in the market.

In crops grown for domestic consumption, the result is not necessarily serious since the affected skin can often be

removed and the undamaged interior can be used. For crops intended for a commercial market, qualitative losses usually result in financial loss. In yams, sweet potatoes and potatoes, diseases causing internal blemishes also reduce the final quality of the crop.

Attacks by Pests

Post-harvest and storage losses are caused by pests, which include: insects, nematodes and animals.

Insects Damage Losses

Insects damage roots and tubers in two ways:

- by boring holes in the tubers, reducing the quantity and quality of the produce and sometimes the germination capacity;
- by damage to the epidermis providing entry for moulds and bacteria to penetrate the tuber.

Estimates of storage losses of roots and tubers due to insects are very scarce. Some investigations in Côte d'Ivoire on damage to stored yams suggest that, under certain circumstances, insect damage losses may reach 25% in just four months of storage.

Insect pests of yams

The following are the most prevalent insect pests; there are others that are of smaller economic significance.

- Pyralid moth (Euzopherodes vapidella). First reported during the 1970s causing extensive damage to stored yams in Nigeria and Côte d'Ivoire. The insect preferentially attacks D. alata, generally during the first few days following harvest. Infestation may also start in the field on those parts of the tuber emerging from the mound. The insect lays its eggs generally on wounds or holes dug by its larvae from a previous generation. The damage is visible from the "dust-like" excrete on the surface of the tuber.

- Tineidae sp. These moths seems to prefer tubers with a lower moisture content which may explain why its attacks generally follow that of E. vapidella on D. alata. The young larvae of Tineidae penetrate the yam tuber from holes previous bored by E. vapidella, remaining unobserved whilst the inside of the tuber is extensively damaged.
- Two species of mealy bugs (Aspidiella hartii) and (Planococcus dioscorea). These are common on D. cayenensis and D. alata. The attack by these pests results in white looking colonies, which can cover the whole tuber, causing mainly a loss in quality and also in reduction in germination.
- The yam beetle (Heteroligus spp.). The beetle attacks the tuber during the growing period damaging the epidermis and permitting the penetration by microorganisms during storage.

Insect pests of sweet potato

Cylas formicarius (F.), widespread through the tropics, and Cylas puncticollis which exists in only a few African countries, are the two most damaging of the many insect pests that attack sweet potato in tropical and subtropical regions. These pests damage the sweet potato roots while the crop is growing as well as during storage. It has been reported that weevil infestation is usually more prevalent in roots harvested during the dry season. In the field the insect lays eggs on the stems and on exposed roots. The larvae, which are responsible for most of the damage, bore holes through the stems and the roots allowing the introduction of other disease organisms.

Insect pests of stored potato

The potato tuber moth (Phthorimaea operculella) is considered the principal insect pest of stored potatoes. In India it is a serious pest in areas where potatoes are stored through "country methods", i.e. not using cool storage. Pest surveys in East Africa carried out by CIP in 1987 showed that the potato tuber moth caused the majority of damage problem.

The moth was also observed to cause extensive damage in stored potatoes on seed farms in Burundi and Rwanda. It is the larval stage that causes the most severe damage. It is about 1 cm in length, has a dark brown (or black) head and a body which may be white, yellow, pink or green. They tunnel extensively into the tuber flesh causing the infected tubers to rot, mainly because of a secondary infection of pathogens

Nematodes

Nematodes mostly infect growing plants and remain on the tubers after harvest. They damage not only the tubers themselves, but also create points of entry for other pests and pathogens. While nematodes are not normally considered of significant importance in stored produce, they have been shown to cause necrotic areas on stored yam tubers. Many species of nematodes affect root and tuber crops. Among these, Scutellonema bradys, Pratylenchus coffeae and Meloidogyne spp are alleged to be the most significant, particularly on yams.

- Scutellonema bradys has been recorded on yams all over the tropics and is recognised as an economically major nematode pest in West Africa. An infestation results in high quantitative losses, as well as loss in the marketable value of the affected tubers. The eggs are most often laid in plant tissues but they are almost always to be found in the soil. Very large populations can build up in the tuber. A tuber disease commonly associated with S. bradys is referred to as "dry rot", where the nematodes feed within the tuber causing tissue breakdown and producing cavities. Initially small yellowish lesions develop beneath the periderm and as the infection spreads these turn into areas of dark-brown or black coloured rot. External symptoms are slight to deep cracks on the tuber skin and malformation of the tuber.
- Pratylenchus coffeae, commonly called "lesion nematode" is found in many countries throughout the world and is known to cause serious damage to a wide

range of crops. It causes a marked reduction in yield and dry rot on stored yams. The symptoms on the affected tubers are similar to that of S. bradys.

— Meloidogyne spp, also known as "root-knot nematode" is very common as root pest of tropical crops often causing serious yield reduction.

The main source of inoculum is reported to be infested seed material. Host populations of S. bradys know a wide range of crops and some weeds. including: cowpea, melon, pigeon peas, okra, tomato, and sorghum.

Attacks by vertebrate pests, even including monkeys, are not very frequent and are not well documented in the literature. When they do happen, the resulting losses may be extensive because of subsequent rotting of the damaged tubers.

Effect of High Storage Temperature

Roots and tubers are susceptible to extremely low or high storage temperatures. Yams, cassava and sweet potatoes are known to suffer from chilling damage at 12°C or below, while for potatoes, cocoyams and sweet potatoes this type of damage occurs at 2°C or below. The extent of chilling damage usually depends on a time/temperature interaction. The most common symptoms are internal tissue breakdown, increased water loss, susceptibility to decay, failure to sprout and changes in culinary qualities, cooking and taste. Potatoes respire strongly at temperatures of 30°C and above.

At these high temperature levels the tubers require a great deal of oxygen and respire a considerable volume of carbon dioxide. At a certain point, the rate of respiration is so great that the cells can no longer obtain sufficient oxygen to sustain the rate induced by the temperature and the carbon dioxide formed cannot be disseminated. This ultimately leads to the death of the cells, giving rise to what is commonly known as "black heart".

References

Bruinsma, D.H., Witsenburg, W.W., Wurden, W., *Selection of technology for food processing in developing countries*, Publ of the International Course in Food Science and Nutrition, Centre for Agricultural Publishing and Documentation, Wageningen, The Netherlands, 1983.

Burton, W.G., *The potato*, A survey of its history and of factors influencing its yield, nutritive value, quality and storage, edit. H. Veenman & Zonen N.V., Wageningen, The Netherlands, 1966.

Diop, A and Wanzie, R., *Improved low-cost storage of potatoes in the North-west province of Cameroon*, Proceeding of the Workshop on the African Experience on Post-harvest Technology Development, Accra (Ghana), 4-8 July 1994.

Hunt, G.L.T., *Low cost potato storage for the tropics*, Acta Horticulturae 158, Tropical Horticulture 10, 1985.

Nahdy, S. M. and Odong, M., " Storage of fresh cassava tuber in plant based storage media", In: Proceeding of the Workshop on *Post-harvest Technology Experience in Africa*, Accra, 4-8 July 1994.

Pushkarnath, *Potato in sub-tropics*, Editor: Orient Longman, 1976.

Rastovski, A., Van Es, A. *et al.*, *Storage of potatoes: post-harvest behaviour, store design, storage practice, handling*, Centre for Agricultural Publishing and Documentation, Wageningen, 1981.

Shaw, R. and Booth, R. (undated), *Simple Processing of Dehydrated Potatoes and Potato Starch*, International Potato Centre (CIP), Lima, Peru, 1981.

14

Potato Processing

Potato is probably the most popular food item in the Indian diet. It is grown all over the country with Uttar Pradesh growing the maximum quantity. Potato is a very rich source of starch. It also contains phosphorus, calcium, iron and some vitamins. Apart from use of fresh potatoes for the purpose of making vegetables and gravy, they are dehydrated in the forms of slices, sticks, cubes or powder to impart better shelf life. Yet another popular use is to make wafers or chips.

Potatoes are grown and used in the Indian culinary since centuries with many end-uses as explained above. However, this note deals with making of potato chips as this product is very popular all over the country and can be manufactured even on a small scale. Potatoes are grown in many parts of the country and thus this is not a location-specific product. This note considers Manipur as the proposed location.

Provisions under the PFA Act must be adhered to. BIS has specified quality standards vide IS 2397:1988

Established and planned potato processing projects aim at solving a very clearly identified development problem. This is that due to insufficient demand, weak infrastructure, poor transportation and perishable nature of the crops, the grower sustains substantial losses. During the post-harvest glut, the loss is considerable and often some of the produce has to be fed to animals or allowed to rot.

Even established potato canning factories or small/ medium scale processing centres suffer huge loss due to erratic supplies.

The grower may like to sell his produce in the open market directly to the consumer, or the produce may not be of high enough quality to process even though it might be good enough for the table. This means that processing capacities will be seriously underexploited.

The main objective of potato processing is to supply wholesome, safe, nutritious and acceptable food to consumers throughout the year. The potato processing activities have been set up, or have to be established in developing countries for one or other of the following reasons:

- diversification of the economy, in order to reduce present dependence on one export commodity;
- government industrialisation policy;
- reduction of imports and meeting export demands;
- stimulate agricultural production by obtaining marketable products;
- generate both rural and urban employment;
- reduce potato losses;
- improve farmers' nutrition by allowing them to consume their own processed potatoes during the off-season;
- generate new sources of income for farmers/artisans;
- develop new value-added products.

Potato Processing planning

The secret of a well planned potato processing centre is that it must be designed to operate for as many months of the year as possible. This means the facilities, the buildings, the material handling and the equipment itself must be interlinked and coordinated properly to allow as many products as possible to be handled at the same time, and yet the

equipment must be versatile enough to be able to handle many products without major alterations.

Advanced planning is necessary to process a large range of products in varied weather and temperature conditions, each requiring a special set of manufacturing and packaging formulae.

The end result of the efforts should be a well-managed processing unit with lower initial investment. A unit which is sensibly laid out and where one requirement co-relates to another, with a sound costing analysis, leads to an integrated operation.

Instead of over-sophisticated machinery, a sensible simple processing unit may be required when planned production is not very large and is geared mainly to meet the demand of the domestic market.

The basic objective is to choose the location which minimises the average production cost, including transport and handling. It is an advantage, all other things being equal, to locate a processing unit near the fresh raw material supply.

It is a necessity for proper handling of the perishable raw materials, it allows the processing unit to allow the product to reach its best stage of maturation and lessens injury from handling and deterioration from changes during long transportation after harvesting.

An adequate supply of good water, availability of manpower, proximity to rail or road transport facilities and adequate markets are other important requirements.

Processing Systems

Small-Scale Processing: This is done by small-scale farmers for personal subsistence or for sale in nearby markets. In this system, processing requires little investment: however, it is time consuming and tedious. Until recently, small-scale processing satisfied the needs of rural and urban populations. However, with the rising rates of population and urbanisation growth and their more diversified food demands, there is need for more processed and diversified types of food.

Intermediate-Scale Processing: In this scale of processing, a group (small-scale processors pool their resources. This can also be done by individuals. Processing is based on the technology used by small-scale processors with differences in the type and capacity of equipment used.

The raw materials are usually grown by the processors themselves or are purchased on contract from other farmers. These operations are usually located on the production site of in order to assure raw materials availability and reduce cost of transport. This system of processing can provide quantities of processed products to urban areas.

Large-Scale Processing: Processing in this system is highly mechanised and requires a substantial supply of raw materials for economical operation. This system requires a large capital investment and high technical and managerial skills. Because of the high demand for foods in recent years many large-scale factories were established in developing countries.

Some succeeded, but the majority failed, especially in West Africa. Most of the failures were related to high labour inputs and relatively high cost, lack of managerial skills, high cost and supply instability of raw materials and changing governmental policies.

Perhaps the most important reason for failure was lack of adequate quantity and regularity of raw material supply to factories. Despite the failure of these commercial operations, they should be able to succeed with better planning and management, along with the undertaking of more in-depth feasibility studies.

It can be concluded that all three types of processing systems have a place in developing countries to complement crop production to meet food demand. Historically, however, small and intermediate scale processing proved to be more successful than large-scale processing in developing countries.

Processing Technologies for Developing Countries

The basis for choosing a processing technology for developing countries ought to be to combine labour, material resources

and capital so that not only the type and quantity of goods and services produced are taken into account, but also the distribution of their benefits and the prospects of overall growth. These should include:

— increasing farmer/artisan income by the full utilisation of available indigenous raw material and local manufacturing of part or all processing equipment;

— cutting production costs by better utilisation of local natural resources (solar energy) and reducing transport costs;

— generating and distributing income by decent-ralising processing activities and involving different beneficiaries in processing activities (investors, newly employed, farmers and small-scale industry);

— maximising national output by reducing capital expenditure and royalty payments, more effectively developing balance-of-payments deficits through minimising imports (equipment, packing material, additives), and maximising export-oriented production;

— maximising availability of consumer goods by maximisation of high-quality, standard processed produce for internal and export markets, reducing post-harvest losses, giving added value to indigenous crops and increasing the volume and quality of agricultural output.

Knowledge and control of the means of production, local manufacturing of processing equipment and development of appropriate/new technologies and more suitable raw material for processing must all be better researched. Decentralisation of activities must be maintained and coordinated.

The introduction of more sophisticated processing equipment and packaging material must be subordinated to internal and export marketing references.

Choosing a technology solely to maximise profits can actually work against true development. Choice should also

be based on a solid, long-term market opportunity to ensure viability. The internal market should be given greater consideration, safeguarded and supported. Training courses, at all levels, in processing and preservation of indigenous crops, must be expanded.

Marketing Potentail

- *Demand and Supply:* Rapid urbanisation and improving standards of living have seen manifold increase in the demand of potato chips. Easy availability, convenient packaging, affordable prices and nutritious values are some other reasons for their popularity even in far flung rural areas. There exists consumer as well as bulk markets.
- *Marketing Strategy:* There are some international as well as national brands but majority of the market is penetrated by local manufacturers due to competitive pricing, easy availability at many outlets and very efficient and timely supply as well as replacement. With proper strategy and network, it is possible to penetrate the market.

Manufacturing Process

Fully grown and ripe potatoes are thoroughly washed before peeling them. Then these potatoes are trimmed and put in brine water for 30-35 minutes to prevent browning. They are afterwards cut in the required sizes on slicing machine.

These slices are blanched in boiling water and are then placed on drying trays which are then put in the drying machine. O Temperature of dryer is maintained in the range of 140 to 150 F. After drying, they are fried in edible oil to make them crisp and brown and then they are packed in polythene bags. The chips could be salty or spicy. Some other flavours which are locally popular can also be tried.

Washing and Peeling

⇩

Brining

⇩

Slicing and Blanching

⇩

Drying and Frying

⇩

Packing

Process Flow Chart

CAPITAL INPUTES

Land and Building

Land of 200 sq.mtrs. with built-up area of 100 sq.mtrs. would be adequate. Equipments would occupy around 60 sq.mtrs and rest of the area can be utilised for storage, a small office etc. The land would cost Rs. 600,000/- whereas cost of building is estimated to be Rs. 7.50 lacs.

Plant and Machinery

Keeping in mind the potential market and economic viability, it is suggested to install production capacity of 40 tons of

potato chips per year with 2 shift working and 300 working days each year. To achieve this capacity, following equipments are required:

Table 1.

Item	Qty.	Price (Rs.)
Slicer made of SS with attachments and electric motor	1	30,000
Electrically-operated dryer with trolleys and 96 trays	1	80,000
Coal-fired Furnace	1	15,000
Motorised Potato Peeling Machine	1	15,000
Automatic Sealing Machine	1	10,000
Cutting and peeling knives, aluminium utensils, weighing scale, etc.	—	30,000
	Total	1,80,000

Miscellaneous Assets

A provision of Rs. 35,000/- would take care of furniture, storage facilities, tables, exhaust fans, etc.

Utilities

Power requirement shall be 10 HP whereas per day water requirement will be 2,000 ltrs. Coal requirement for furnace shall be about 1 ton every month.

Raw Material

The all-important raw material is good quality potatoes. Potatoes have high water content. Hence total process loss shall be almost 30%. The state of Manipur produces around 70,000 tonnes of potatoes every year. Thus, procurement of the required quantity shall not be a bottleneck. Other materials like salt, spices, preservatives etc. are easily available.

Proposed Project

— *Land and Building:* Land of about 200 sq.mtrs. with built-up area of 100 sq.mtrs. would suffice. Land may cost

about Rs. 600,000/- whereas the cost of construction would be in the vicinity of Rs. 7,50,000/-.

— *Plant and Machinery:* The total cost under this head is estimated to be Rs. 1.80 lacs as explained earlier.

— *Miscellaneous Assets:* An estimated expenditure under this head is Rs. 35,000/-.

— *Preliminary & Pre-operative Expenses:* Expenditure like registration and administrative charges, pre-production expenses, trial run expenses etc. is estimated to be Rs. 35,000/-.

— *Working Capital Requirement:* At 60% capacity utilisation in the first year, the total working capital needs will be Rs. 1.36 lacs consisting of bank finance of Rs. 0.87 lacs and margin of Rs. 0.49 lacs. The detailed calculations are as under: Financial assistance in the form of grant is available from the Ministry of Food Processing Industries, Govt. of India, towards expenditure on technical civil works and plant and machinery for eligible projects subject to certain terms and conditions.

Table 2. Manpower requirements

Particulars	Nos.	Monthly Salary (Rs.)	Total Monthly Salary (Rs.)
Skilled Workers	2	1,800	3,600
Semi-skilled Workers	2	1,500	3,000
Helpers	2	1,200	2,400
Salesmen	1	1,500	1,500
		Total	10,500

Profitability Calculations

— *Production Capacity and Build-up:* The rated production capacity of the plant would be 40 tonnes per year and it is anticipated that it shall be operated at 60% and 75% respectively during first two years.

— *Sales Revenue at 100%:* For a new entrant, it becomes very difficult to capture the market. Hence, it is

assumed that the net selling price will be Rs.55/- per kg. or Rs. 55,000/- per ton. Thus, the total sales realisation will be Rs. 22.00 lacs.

— *Utilities:* Annual cost under this head at 100% activity level shall be Rs.50, 000/-.

— *Selling Expenses:* Since the unit will be entering the competitive market, it is necessary to undertake advertisements in local media. There will be other costs like transportation and selling commission. Hence, a provision of 25% of sales value is adequate every year.

— *Interest:* Interest on term loan of Rs. 4.70 lacs is calculated @ 12% per annum assuming repayment in 3 years including a moratorium period of 6 months whereas on bank assistance for working capital, it is calculated @ 14% per annum.

— *Depreciation:* It is computed on WDV basis and rates assumed are 10% on building and 20% on plant and machinery.

REFERECES

Biaugeaud, H., *Food processing equipment,* Technical Bulletin, Henri Biaugeaud, S.A.: Arcueil, 1994.

Colin, D., "Recent trends in fruit and vegetable processing", *In Food Science and Technology Today,* 1992.

Magee, T.R.A. and Wilkinson, C.P.D., "Influence of process variables on the drying of potato slices", *In Int. Journal of Food Science and Technology,* 1992.

Matz, S.A., "Potato Chips", *In Snack Food Techology,* AVI Publishing Co., Inc. Westport, Conn., 1984.

Werge RW, "Potato Processing in the Central Highlands of Peru", *Ecology of Food and Nutrition,* food processing, potato; South America; Peru, 1979.

15

Marketing of Potatoes

Marketing is a series of interconnected activities. In the case of horticultural marketing these include:

- planning production;
- growing and harvesting;
- grading of products and their packing, transport, storage, processing, distribution and sale;
- sending information from production area to market and from market back to producing areas.

All of these activities are links in the production-marketing chain. Like any chain, it is only as good as its weakest link. Marketing systems are dynamic. They are competitive and involve continuous change and improvement. Suppliers who have lower costs, are more efficient and can deliver quality potatoes are those who survive and prosper. Those who have high costs, do not adapt to changes in market demand and provide poorer quality are often forced out of business.

Potatoes are produced seasonally, but the market requires products throughout the year. For many decades, this problem of matching product availability with consumer demand was solved by two methods: (1) selling fresh products during harvest and shortly thereafter, and (2) processing the rest to meet demand during the remainder of the year. As technology improved and consumer incomes increased, it became possible to provide fresh produce year-round. Consumers now expect fresh tomatoes, strawberries, and

sweet corn every month of the year. In addition, a strong demand for processed fruits and vegetables remains.

Increased consumer incomes and year-round demand for fresh produce force retailers or their representatives to establish buying points both in different climatic regions. Some retailers make year-round contracts with fresh protato packers, who may contract with growers. The contracts and large-volume buying practices enable packers to obtain sufficient quantities of individual products.

Large potato packers may contract with growers in several different production regions to ensure that fresh fruits and vegetables are available every week of the year. These packers contract only in regions with a large number of growers. Further, they contract mainly with the largest growers, even in concentrated production regions. Some packers ensure supplies by growing commodities themselves. Large retailers and packers are unlikely to purchase products directly from a single, small-scale grower, especially a grower in a remote production area.

To stay competitive, potato processors must keep unit costs low, so large corporations generally do the processing. Many of the smaller processors have found it difficult to compete with these corporations and have gone out of business. Almost all processors contract with growers for their raw-product needs. Over the past 20 years, most processors have moved their plants to major production areas. It may be difficult to locate processors interested in buying from small-scale growers. If a processor does indicate a willingness to buy, insist on a purchase contract with the processor before planting the crop.

You may wish to process your own fruits and vegetables. This is how many of today's processors entered the business several decades ago. If you are considering a processing operation, contract your state department of agriculture to learn more about food regulations pertaining to the processing, packaging, and labeling of your product. It will probably be impossible for you to process and market produce at costs as low as those incurred by larger processors. As a

result, it may be very difficult to compete for sales with your processed fruits and vegetables through conventional retail outlets. Selling processed fruits or vegetables may, in addition, increase your exposure to product liability suits. Sometimes a small processor will process potatoes for a grower on a contract basis, especially if the product being produced does not interfere with their own processing requirements. Of course, their busy time is often when you need to process your product, so this works best for something with an especially early or late processing season.

Market Demand

Larger potato growers, particularly those located in major production areas, can pursue either of the two traditional marketing alternatives: wholesale fresh marketing or processing. Small-scale growers who find these marketing avenues closed to them will need to take a direct-to-consumer approach. This requires thorough research of the market and customer behavior before planning crop production. Some farmers generate profits by planting and then looking for a market, but this is extremely risky for protato growers. There are far more failures than success stories. If you are a new grower, or an established one planning to produce a new item, you should first attempt to evaluate the market demand for the product and then decide which direct marketing channel(s) will best meet the needs of your consumers. Your estimates of profitability should include the marketing channel costs as well as production costs.

Small-scale growers should collect three types of information before deciding to produce and market fresh fruits and vegetables.

— *Determine and define the geographic area where you will market potatoes.* Identify potential customers before you investigate consumer demand.

— *The level of unfulfilled demand among consumers within the defined marketing area.* It is advisable to estimate the amount that consumers (buyers) within that market

presently buy. In the process, you will gain insight into how they might be better served.

— *Consider the competitive structure of the market.* Knowing who your potential competitors are, where they are located, and what services they provide are important pieces of information for you as a new grower-marketer. Note potential competitors who might have marketing advantages (lower costs, better locations, and higher-quality produce) or may provide potential consumers with similar products.

Cooperative extension personnel are a valuable source of information about the local marketing situation. In addition, cooperative extension offices often provide training workshops and seminars specifically for direct marketers. Determine the likely impact of increased production on future selling prices. If you place more produce on the market, and the products are not of different quality or do not meet some other "unmet need" for which consumers are willing to pay a higher price, then it is likely that prices will drop from current levels. An expected price is a vital piece of information for planning purposes. There is no simple, reliable way of predicting local market prices, but such information is very important to growers. Estimate prices by considering all the available information and by using good judgment. When using these estimated prices for planning, remember to include marketing costs and the cost of unsold product.

It is often difficult to introduce a new product to consumers and get them to buy it, because they may not be familiar with it or its potential uses. The learning process takes time. A new item may provide you with a foothold in a market that will grow rapidly. However, it is more likely to grow slowly, which may result in product waste during the first few years. If fruits or vegetables are commonly consumed, you must find out whether buyers will increase purchases when more is grown and sold in the manner you choose.

Supermarkets are major suppliers of potatoes, but generally are not strong competitors for sales of in-season

crops. For example, supermarkets sell very little sweet corn when it is available at farmers markets or along roadsides. Other direct marketers, whether they are pick-your-own operations, roadside markets, farmers markets, or curb markets, are strong competitors, and you should note their presence. In recent years, it is not uncommon for direct marketers to express concern about the increased number of competitors and the possibility of profit loss in existing operations.

Direct Marketing

Several factors must be considered when choosing a direct marketing channel. Location can have a major impact on an enterprise's profitability because location affects the direct marketing channel used as well as the ability to attract customers.

Bringing potatoes to market requires special handling. To preserve quality and maintain marketability, each crop must be harvested, prepared for market, attired, and shipped. Four different direct marketing channels are used by small-scale growers. These channels vary in the amount of labor and capital the marketer must provide and in the location of the marketplace.

- Subscription farming is essentially grower contracting. You "contract" directly with the consumer to produce and deliver a specific vegetable or fruit. This system may require maintenance of a delivery vehicle, good consumer relations, and a wide variety of fruits and vegetables. This approach may be a good way to profit from special attributes of your produce, such as products with ethnic appeal, organically grown produce, and gourmet items.
- Pick-your-own, sometimes referred to as U-pick, operations require the least grower labor and capital for market facilities. Your customers perform a large amount of the marketing function and do the

harvesting. This method works well for some commodities and in some locations, but not for all crops or for all growers. It usually works best for potatoes that are purchased in fairly large quantities for home processing.

— Roadside marketing is by far the most common direct marketing system. The system requires some capital investment in a facility, and you have to provide the harvesting, market preparation, and retailing labor. Customer relationships are extremely important, but they are more conventional than the relationship between a clerk and a customer.

— Farmers markets or "curb markets" are similar to roadside markets, but the retailing function is simply moved closer to the consumer. This enables you to offset the disadvantages of your production location. One advantage is generating high levels of customer traffic. Staffing needs are simple to plan because operations generally occur only during specified hours. One big disadvantage, however, is the need to predict sales so that enough produce can be harvested and prepared each market day. At a farmers market, you generally cannot replenish stock by quickly harvesting more, which can sometimes be done at roadside stands. Also, a rainy farmers market day can result in unsold stock that cannot be sold on the next market day.

Price and quality are synonymous in potato production. Unfortunately, it is not always easy to know what is meant by "high quality" and quality judgment often varies from year to year. Buyers and consumers, however, often have additional criteria by which they judge produce quality, including ripeness, cleanliness and the presence of insects and foreign material.

Proper disease management, harvest practices, and postharvest handling are critical to marketing success. Cooling produce to remove field heat and improve shelf life is especially important. Treatments to reduce decay may be another important consideration. Sorting and washing of

some potato can also be done to help maintain quality and improve appearance. For certain crops like small fruits and other delicate produce, sorting and/or washing is not an option; harvest crews must be well trained and quality continuously monitored to assure a marketable crop.

Effect of Improved Marketing Systems

Most potato farmers see themselves as "price takers", thinking that they have no control over prices and have to accept what is offered. They do not always know how to find new buyers nor how market demand is changing and which potatoes are most profitable to grow. They lack the understanding to improve the prices they receive and the profitability of their production.

Your role as an extension advisor should be to help potato farmers become better informed about the market. They can then start to make decisions on how to improve their marketing. However, you must never tell farmers what they should do or what products to sell. Commercial decisions like these must be made by the potato farmers themselves. There are two reasons for this:

— Potato farmers need to become owners of new ideas. In this way they become committed and aware that they are responsible for the success or failure of what they do. This increases the chances of a successful outcome.

— If an extension officer (or NGO specialist) recommends a course of action that fails, he or she will be blamed by the potato farmers.

When asked about their problems, potato farmers commonly identify marketing issues as their key constraint. Problems highlighted are usually lack of markets, poor prices, inadequate roads and poor communications. However, while potato farmers can usually state clearly their problems they often face difficulties in identifying potential solutions. This guide aims to help you to help farmers work out what to do. Potato farmers are generally highly skilled in agricultural techniques but marketing requires learning new skills, new

techniques and new sources of information. Armed with business and marketing skills farmers will be better able to run their farms profitably.

Small-scale potato farmers face the biggest marketing problems. Compares the strengths and weaknesses of large-scale farms and small-scale farms. Small farmers need the most support and that their success depends on getting the best prices possible. This can be done by obtaining better information about marketing and the different marketing options available to them.

Strengths and Weaknesses of Small and Large Potato Farms

Small Potato Farms

Strengths: Cheap family labour is available. Smalls potato farms are suitable for labour-intensive products (e.g. those requiring transplanting, pruning and multiple harvests by hand). Small potato farms can grow products that require attention to detail. Small potato farms can effectively supply low-volume specialised niche markets and value-added products, such as herbs, flowers and ornamental plants. They can also supply local processors.

Weaknesses: Small farms produce only limited quantities. They are often located far away from major markets. Education standards are often low. Small farmers are reluctant to introduce new technology. They face difficulties in obtaining information, capital and support. They are weak in negotiation and often lack confidence, especially when dealing with traders and companies. Small farmers tend to be averse to risk. They need income stability and cannot afford losses.

Large Potato Farms

Strengths: Large potato farms are suitable for mechanised, large-scale production for major crops like wheat, sugar cane and maize. They can grow crops that require a large capital investment. They are best equipped to produce and sell

produce in large volumes to major buyers. They have access to capital, information and technology.

Weaknesses: Large farmers often have high overhead costs. They can be poor at organising and controlling large numbers of workers. They cannot easily service small and niche markets. In some countries they may be criticised for using land that could be used by poor farmers.

Role of Marketing

Marketing and the Rural Economy

Those who carry out marketing have a strong incentive to increase the value of rural trade, because increased sales should lead to higher profits. Rural businesses include suppliers of inputs, buyers of produce, transporters, storage companies, processors and wholesalers. They can range in size from individual entrepreneurs to large-scale agribusinesses, but whatever their size, all stand to gain from improvements in the marketing process.

Businesses are often said to be exploiting potato farmers and making unfair profits. They certainly try to maximize their profits, but without such businesses farmers would not be linked to markets and would not be able to sell all their produce. Traders and other rural businesses can help farmers find new markets and lower their costs. All of this leads to improved production opportunities and higher incomes for farmers. Rural marketing businesses are often small, have limited resources and are traditional in outlook. Identifying new markets, advising on technologies and improving understanding of markets are all ways in which extension workers can help traders and potato farmers to increase sales.

Governments can help farmers in many ways, without actually working with them directly. Promotion of competition, provision of market information and improvement of market infrastructure are powerful ways to ensure good returns for potato farmers.

Marketing and Consumers

Consumers want to pay low prices. Potato farmers want to receive high prices and to be paid as much of the consumer price as possible. The best way of achieving a balance between these two conflicting aims is through an efficient and low-cost marketing chain. This generally involves using larger scale transport, reducing losses, and reducing other costs.

Consumer preferences for food products are constantly changing and developing, particularly in the case of horticultural products. Consumers need a production-marketing chain that can respond to their changing tastes. The marketing system needs to deliver the volumes, quality and variety of safe and nutritious food products that consumers require. It needs to be sufficiently dynamic so that it can continue to provide consumers with choice by developing and delivering new products.

Helping Producers to Farm Profitably

In the past, ministries of agriculture and even NGOs worked mainly on improving agricultural production and yields. This involved carrying out field trials, testing varieties and developing improved production techniques. The extension officer's primary role was to advise and encourage farmers to use improved production technologies. Although potato farmers do, of course, benefit from improved yields, their primary concern must be to make a reasonable living from their potato farm. They need to generate sufficient profit from their sales to be able to cover their living costs. The key issue for them is how to do this.

The main role of the extension officers should therefore be to help farmers farm more profitably. To do this they must have a basic understanding of the factors that influence profits. These are farm gate or market prices, volumes sold and costs.

Balancing Risk and Profit

High profits are closely linked to high risks. In agriculture, the

enterprises that offer the possibility of the highest profit are usually the most risky. Some products may be technically difficult to produce or susceptible to pests and diseases. Farmers, and particularly poorer farmers, need to consider the importance of balancing production of potentially more profitable, but riskier, crops with those that provide a lower, but more stable and reliable income. Enterprises that usually provide reliable incomes include staple crops and contract production for agricultural processors.

Potato Farmer Prices

The prices that farmers receive are mostly influenced by supply and demand.

Other important factors affecting the price a farmer receives include:

— *How much competition there is between buyers:* When there is only one buyer, he or she will possibly offer a low "take it or leave it" price. Prices are likely to be higher and more profitable when there are many buyers competing with each other.

— *The amount of information the farmer has:* A farmer who is poorly informed of market prices and demand will be less able to negotiate with traders.

— *The quality of the produce:* Buyers will sometimes, but not always, offer higher prices to producers with better quality potatoes. If most farmers produce low quality produce it is difficult for a trader to market high-quality produce separately.

— *The transport costs:* Lower prices are generally offered to producers whose product is costly to transport Marketing.

Price Increases along the Marketing Chain

As produce moves along the production-marketing chain, prices increase. Unit prices are lowest when farmers sell a standing crop. An example of this is when fruit growers sell their fruit on the tree to fruit contractors, who then harvest and

pack it. This is common in Near East and South Asian countries. Harvested produce sold at the farm gate to rural buyers obtains a lower price than produce sold at a local assembly market which, in turn, is sold at a lower price than produce sold at a wholesale market. Retail sales achieve the highest price. The price the farmer gets depends on the point in the marketing chain at which he or she decides to sell.

Although prices are higher, selling further along the marketing chain involves additional costs for transport, market fees, meals and accommodation. There are also costs in terms of farmers' time. Farmers who take produce to market and sell directly to consumers will usually get the highest price but they do need to decide whether this is the best use of their time, as it may be more usefully spent managing the farm.

Farmers' Marketing Costs

Marketing costs can be reduced, especially by achieving economies of scale. The larger the volume of product marketed, the greater the scope for lowering unit costs, for example through using larger vehicles or bulk-buying of packaging. Sometimes, traders operating on a commission basic will accept lower percentage commissions for large quantities. Small farmers working on their own are naturally at a disadvantage. Cost savings have sometimes been achieved by farmers working as a group to jointly purchase inputs and hire transport.

Marketing Costs and the Trader

Traders also have costs. Because these are not always visible, traders are often accused of making unreasonable profits. Although traders can sometimes exploit a situation and make exceptional profits, there are also occasions when they are unable to sell potatoes that they have purchased and therefore make losses. High profits on some occasions compensate for risks of losing money on others.

Product Losses

Losses of potatoes will inevitably occur during marketing and are one example of a marketing cost that is often invisible. Types of loss are:

— *Loss of weight:* Produce can lose weight, mainly through water loss in storage, transport and marketing. A trader may buy a tonne of potatoes, but as a result of water loss may only sell 975 kilos.

— *Loss of value over time:* The fresher the product, the higher the price. Fresh produce will last longer, will probably taste better and may be more nutritious. Over time, bruises and pest damage become more apparent and water loss leads to many products starting to shrivel, all leading to losses in quality and in price. The most obvious loss occurs when produce has to be thrown away.

— *Physical losses:* Fresh produce losses in developing countries can be exceptionally high. Losses of ten to thirty percent are often quoted. In practice, only on rare occasions is this volume of produce thrown away. These figures more accurately reflect the loss in value that occurs from the combination of moisture loss, damage, disease and loss of freshness, as well as the proportion of the harvest that is thrown away.

For a trader, minimizing losses is very important. Part of the difference between the consumer price and the price paid to the farmer is because the trader needs to make a large margin, to cover wasted or damaged produce. By only looking at the selling price it is easy to think that the trader is making an unfair profit. However, when losses are taken into account the picture changes.

Supply and Demand

The quantity of produce that consumers want to purchase is affected by many factors, the most important being:

— price of the goods;

— tastes and preferences of the consumers;
— number of consumers;
— incomes of consumers;
— prices of competing produce;
— range of products available to consumers.

The quantity that producers supply is also affected by a number of factors, the most important being:

— price of the goods/products on the market;
— price of inputs/costs of production;
— technological factors;
— climate;
— storage possibilities.

The price of a product is mainly determined by supply and demand. Basically, a balance is achieved between what people are prepared to supply at a price and what people are willing to pay for the product. As the price rises the quantity that will be supplied also rises and the quantity demanded falls, and vice versa. It is important to note that:

— Supply is what producers are prepared to sell at a certain price
— Demand is how much consumers are prepared to buy at the market price

While supply is influenced by production it is not always the same as production. For less perishable crops, farmers or traders may decide to store them in the hope that prices will rise, rather than sell them immediately. When prices do rise they may take the products out of store to sell. At this time supply is equal to production harvested for immediate sale plus products taken out of store. However, it should be stressed that demand is not how much people would like to buy, nor what they should buy for a healthy lifestyle. It is what they are prepared to buy at the prevailing price.

Short-Term Price Fluctuations

Unlike the prices of staple foods such as maize or rice,

horticultural product prices fluctuate enormously. They can fluctuate from day to day and during the day, depending on supply and demand.

The main causes of short-term price changes of fresh produce are:

- The amount of produce on sale in the market on a particular day and the quantities sold in the previous few days.
- Short-term demand changes.
- The effect on demand of prices of competing products.

To take advantage of opportunities when prices are high a supplier needs to be in close communication with the markets and able to transport produce rapidly. In many countries farmers are increasingly able to deliver products, such as leafy vegetables, to markets when prices are high, using mobile telephones to contact buyers.

Seasonal Price Changes

In countries with pronounced seasons, supplies are low at the start of the harvest season, so prices are high. Prices are at their lowest when the crop reaches maturity in the main production areas. At the end of the season prices normally increase again as supply diminishes. Prices are generally highest during the off-season, when only a small percentage of farmers are able to grow the crop. Farmers who are located in areas where early or late-season crop production is possible or who can use production methods, such as plastic tunnels or greenhouses, that bring forward the harvesting date, are best placed to take advantage of high early or late-season prices.

Long-Term Balance between Supply and Demand

The theoretical point at which supply and demand are in balance is referred to by economists as "equilibrium". At this price there is enough incentive for farmers to produce the quantity that consumers will buy at that price. In practice, although the marketing chain tries to achieve this balance it is rarely done because there are so many factors affecting both

supply and demand and because farmers lack adequate information about demand.

References

Brook, R. C., R. J. Fick, and T. D. Forbush, *Potato Storage Design and Managemen*, East Lansing, MI.: Michigan State University Agricultural Engineering Information Series, AEIS#627, 1995.

Christ, B. J., *Potato Diseases in Pennsylvania,* Pennsylvania College of Agricultural Sciences, University Park, PA. 1998.

16

Potato Supply Chain

The potato industry has operated under free market conditions for many years. As a result, the deregulation process of the agricultural sector during the 1990's did not have a great impact on the potato industry. The South African potato crop is marketed as seed potatoes, table potatoes and potatoes earmarked for processing. Table potatoes sold on fresh markets and directly to consumers comprise the bulk of the produce, as indicated in Table 1.

Table 1. Utilisation of the South African potato crop

	1995		*2001*	
Utilisation	*Production (Ton)*	Percentage	Production (Ton)	Percentage
Table (Local consumption)	1,065,960	71.8	1, 017,114	63.5
Seed	180,139	12.1	208,477	13.0
Processing	152,770	10.3	270,445	16.9
Exports	86,000	5.8	106,000	6.6
Total	1,484,869	100	1,602,036	100

Figure 1is an illustration of the different marketing channels for table potatoes in South Africa.

Informal trading creates an enormous welfare effect, because the bulk of these potatoes are pre-packed and distributed further in city centres and townships. The 35,2%

potatoes not marketed on the fresh produce markets are designated either for export, processing or direct trade.

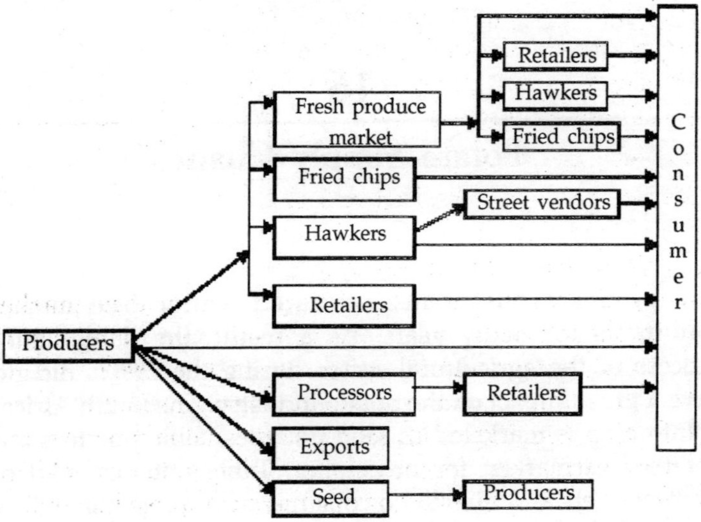

Figure 1. Marketing channels for table potatoes

Table Potatoes

The various fresh produce markets and the volume of table potatoes traded on these markets for the 2001/2 and 2002/3 seasons. It is significant that the turnover increased with an average of 42% and the supply decreased with an average of 16% between the two seasons.

Seed Potatoes

There are approximately 400 seed growers in South Africa, who produce seed potatoes throughout the year. The certified yield per hectare has increased constantly over the past few years.

Registered seed potato production is executed on 8,000 to 10,000 ha, annually, in six main areas. The annual certified yield amounts from 4,000,000 to 5,200,000 x 25 kg bags. Seed

potato production, therefore, comprises approximately 13% of the total potato production. During 2001, approximately 82% of the total potato crop was planted with certified seed in comparison to 74% during 1995 and 50% during 1985.

Table 2. Seed potato production

Season	Registered plantings (Hectares)	Certified yield (25 kg bags)
1999/2000	9 637,89	5 162 691
2000/2001	9 505,16	4 977 970
2001/2002	8 398,81	4 230 954
2002/2003	8 437,05	3 931 849

The import into the country of conventional seed potatoes is not allowed due to the high risk of importing tuber-borne diseases, which will place the local industry at risk. In vitro materials as well as mini-tubers from approved institutions are imported, however, to establish new varieties in South Africa.

Processed Potatoes

The South African potato processing industry grew by more than 100% from 1991 to 1995. This growth took place primarily in the three main disciplines in the processing industry, namely frozen fries, fresh fries and crisps. During the past five years, this growth slowed down, mainly for economic reasons, and also because there was a shortage of good quality potatoes for processing. The growth that took place since 1991 can be ascribed largely to the following factors:

— the changes in economic circumstances
— the expansion of the fast-food industry
— the higher average income of the population
— the rapid rate of urbanisation
— the influx of international processing companies

Mixed vegetables represent approximately 0,78% of the total processing industry in South Africa, and have shown a decline in production since 1997. The cultivars most commonly used for mixed vegetables are Vanderplank, Buffelspoort and BP1. Frozen French fries represent 43,9% of the total processed potato products in South Africa. There has been a large increase in the manufacturing of this product over the last couple of years. The expansion is mainly caused by today's fast-paced life-style.

The growth in production is also the result of expansion of the existing facilities. Fresh French fries represent approximately 16,5% of the total processed potato products in South Africa. The manufacturing of fresh French fries has shown a decrease over the last couple of years. This decrease in production is the result of a decrease in the number of companies involved in the industry, and the strong increase in frozen French fries production, which took over a large part of the market. Only limited quantities of potatoes are canned in South Africa. This canned food is mainly in the form of mixed vegetables, to which potatoes can contribute up to 20% of the mixture. These days skinned baby potatoes are also canned.

Crisps

Crisps represent approximately 37,4% of the total processed potato production in South Africa. The steady growth in production over the past five years is the result of an expansion of the existing factories, and an increase in the number of companies involved in the industry, although certain factories and smaller companies closed down during 2000 and 2001. The cultivars most commonly used for crisps are Hertha, Pimpernel, Lady Rosetta, Fiana, Crebella and Ernstestoltz. Other processed products include inter alia hash browns, potato bites, dehydrated products, baby food, etc. They represent approximately 0,4% of the total processed potato production in South Africa.

Potato Production

Potato production in South Africa is a high yield, high-risk enterprise with good income potential for producers, yet it is an affordable food available to most people (rich and poor).

The most significant changes in the production patterns in potatoes during the past two decades can be attributed to the implementation of irrigation technology and a major shift from dry land production towards irrigation production. High production costs compelled the introduction of irrigation technology by farmers to manage high risks and price fluctuations characteristic of dry land potato production.

Dry land production decreased from 33,543 ha in 1991 to an all-time low of 12,915 ha in 2000, whereas irrigation production increased from 32,251 ha in 1991 to 38,000 ha in 2002/3. In 2002/2003 there was a decline in irrigated hectares planted with potatoes as farmers shifted to maize because of the higher prices of maize.

One of the major risk factors associated with dry land production is "over production" in good years with market prices below production costs, while, "under production" in dry years would result in very good market prices. The resultant fluctuation in quantity and price places a heavy burden on the infrastructure at fresh produce markets and impacts negatively on consumer preferences.

Production Regions

Potatoes are produced all over South Africa in different climatic regions. This results in a continuous supply of potatoes throughout the year. Consequently, domestic consumers have almost continuous access to fresh potatoes. Table 3 clearly illustrates the Eastern Free State as the major dry land producing area with 9,094 ha of dry land potatoes and 1,506 ha potatoes under irrigation. Limpopo with 6,973 ha under irrigation is the region with the most potatoes under irrigation.

Table 3. South African potato production in the 16 regions during 2001

Region	Hectares			Producers	
	Dry land	Irrigation	Total	On list	Planted
Sandveld	-	7,558	7,558	151	143
Ceres	-	1285	1285	53	43
South Western Cape	-	589	589	79	34
Southern Cape	-	472	472	34	20
Northern Cape	-	1,786	1,786	56	22
Eastern Cape	-	2,640	2,640	77	60
North Eastern Cape	1,378	823	2,201	94	60
Western Free State	2,722	3,884	6,606	174	117
South Western Free State	85	1,786	1,871	80	51
Eastern Free State	9,094	1,506	10,600	132	117
KwaZulu-Natal	-	3,483	3,483	103	53
Mpumalanga	822	3,789	4,611	75	73
Marble Hall	-	1,591	1,591	27	24
Limpopo	-	6,973	6,973	95	95
Northwest	-	925	925	67	20
Gauteng	-	595	595	17	15
Total	14,101	39,685	53,786	1314	947

The region with the most producers is the western Free State with 174, followed by the Sandveld and the eastern Free State with 151 and 132 producers, respectively.

Production Costs of Potatoes

Production costs of potatoes vary between R20,915 per ha for dry land production to more than R42,000 for both seed production and the production of table potatoes under irrigation. A breakdown of the average production costs needed for the production of potatoes shows that seed is the largest input component (18,2% of total input costs), followed by transport (10,9%), fertiliser (10,5%), chemicals (10,4%) and packaging (10,3%). An index of the different inputs was used to calculate the respective contributions of the various inputs to the increase in production costs since 1995.

The prices received for potatoes during the third quarter of 2002 exceeded the index prices for inputs during the same period. Average potato prices during that period varied from R27-00 to R31-50 per 10 kg but they decreased to less than R20-00 per 10 kg from January 2003 onwards.

Cost of Fresh Potatoes

It is believed that potato prices are driven by supply and demand with distinctive seasonal trends. According to Potato South Africa, 63% of the potato crop is delivered on the fresh produce markets, and market prices therefore act as a barometer for producer prices in other sectors of the potato industry. Processors entering into contracts with producers confirmed that market prices serve as a major barometer when contract prices with producers are negotiated. It is evident from Table 4, that potato prices and the supply of potatoes on fresh produce markets are negatively correlated.

Potato prices decline with increase in supply, and vice versa. An a priori observation of supply and price trends, indicates a decline in supply and an increase in price during the period January 2000 until July 2003. The main reason for the dramatic increase in price during the third quarter of 2002 was a decline in supply during the months August (4,9 million x 10 kg), September (4,8 million x 10kg) and October 2002 with 5 million x 10 kg.

The total supply of potatoes on the fresh produce markets during those three months adds up to 14,8 million x 10 kg pockets, compared to 21,3 million x 10kg during the same period in 2001 and 19,2 million x 10 kg pockets during that period in 2000.

However, no a priori conclusion can be drawn merely from this graph. Further tests were run, therefore, to confirm the above results. A measuring and comparison of volatility of real potato prices and the exchange rate was done. By comparing volatility calculations for the exchange rate and potato prices, which is time varying, it was concluded that the real price of fresh potatoes was not affected by the exchange rate1.

Table 4. Average production costs for potatoes in South Africa (per ha)

	Commercial production			
Costs	Seed production (R)	Dry land (R)	Irrigation (R)	Processing (R)
Soil preparation, spraying and harvesting	1,420	1,050	1,420	1,420
Seed	8,802	3,815	7,725	7,800
Fuel &lubricants	2,500	1,200	2,500	2,500
Fertilizer	4,500	2,150	4,500	4,590
Water & electricity	1,700		1,700	1,700
Disease-, pest-and weed control	4,500	2,600	4,000	4,200
Repair & maintenance	3,750	1,900	3,750	3,750
Packaging	3,000	1,500	5000	
Marketing	2,700	2,000	3,200	
Transport	4900	2,000	4,900	
Certification & laboratory fees	1,090			
Other	3,700	2,700	3,700	3,700
Total Costs	42,562	20,915	42,395	29,660

Table 5. Number of 10 kg potatoes sold on all fresh produce markets and prices in cents per 10 kg received from Jan 2000 to July 2003

Month	Number (10 kg)	Price (c/10 kg)	Month	Number (10 kg)	Price (c/10kg)
Jan-00	7467279	1324.251	Oct-01	6871046	1375.482
Feb-00	6050465	1314.986	Nov-01	7881758	1767.109
Mar-00	6633031	1093.921	Dec-01	6079257	1872.044
Apr-00	6416453	1184.99	Jan-02	7715441	1352.657
May-00	8556859	1205.743	Feb-02	6591308	1131.168
Jun-00	7282310	973.986	Mar-02	7264869	990.3464
Jul-00	9191980	976.7952	Apr-02	6657229	1205.968
Aug-00	6377551	1272.187	May-02	7999653	1497.728
Sep-00	6380368	1395.115	Jun-02	5751305	1940.724
Oct-00	6440101	1471.047	Jul-02	7076656	2150.526

Nov-00	9327862	1117.567	Aug-02	4979832	2717.245
Dec-00	6700784	1245.649	Sep-02	4895905	3188.858
Jan-01	7861172	1123.359	Oct-02	5012983	3171.388
Feb-01	6630400	1018.974	Nov-02	7435127	2247.744
Mar-01	6899330	977.6518	Dec-02	5770603	2384.029
Apr-01	6904152	1060.756	Jan-03	6523037	1927.426
May-01	9171791	1030.983	Feb-03	5555961	1787.075
Jun-01	7698324	906.6085	Mar-03	5764176	1915.232
Jul-01	9675261	983.2467	Apr-03	5534517	2017.84
Aug-01	7427170	1004.177	May-03	7079028	2181.157
Sep-01	7020636	1196.733	Jun-03	5612483	2069.261
Oct-01	6871046	1375.482	Jul-03	6253923	1833.752

Another important aspect in determining the price trends in the market for fresh potatoes is tracking the margin between the price at the fresh produce market and the average retail price for a 10kg bag of potatoes. Despite the fact that the margin peaked at R20 per bag in January 2003, for the rest of 2003 the margin stayed at a normal R10 per bag. To an outsider it may seem that the January 2003 appointment of the Food Pricing Monitoring Committee had a dramatic impact because retail margins suddenly dropped back to normal levels of R10 per bag.

Cost of Processed Potatoes

The processing industry mainly makes use of forward contracts with potato producers. The processors negotiate product prices with producers and in some instances they supply and support producers to obtain the necessary inputs. Long-term relationships are normally established between the processors and their suppliers. It is, therefore, imperative that both parties experience a win-win situation in contract production. The prices on the fresh produce markets, therefore, serve as an important determinant of contract prices. For many years the gross margin figures for contract farmers have been below the earnings of producers focusing on fresh potato markets, but contracting protects them from the volatility of the prices on the fresh produce markets.

Processors were very hesitant to supply a breakdown of their production costs for fear that their competitors might obtain this information. A breakdown of detailed production costs, however, might shed light on the real reasons for the increase in consumer prices for processed potatoes. The reason for the sudden decline in margins might be attributed to either the impact of the time lag effect on potato producer prices or on processing costs.

References

Jordaan, A.J., and Van Schalkwyk, H.D., *Impact of the Potato Industry on the South African Economy*, Department of Agricultural Economics, Free State University, Bloemfontein, 2002.

Potatoes South Africa, *The South African potato industry. A review*, Potatoes South Africa, Pretoria, 1999.

17

Potato Production Trends

The current trend in potato production is being driven in part by a shift in food habits, as demand for potatoes increases with greater absolute numbers of people with relatively high incomes. The potato is not primarily a rural staple, but a cash crop which provides significant income to rural farmers. The potato processing industry is still undeveloped, but has strong potential to grow in response to domestic and international markets. Current and potential gains in productivity should, however, be considered in light of overall population growth.

Potato Production in India

The early historical record of the potato in India is unclear since the term "potato" is derived from "batata," the Carib term for sweetpotato (Ipomoea batatas), which preceded the potato by eighty years in its introduction to Europe from its site of origin in the Andes of South America.

The earliest known reference to "potato" in India is from an account by Edward Terry, who was chaplain to Sir Thomas Roe, British ambassador to the court of the Mughal Emperor Jahanagir from 1615 to 1619, in the northern area of contemporary India.

In his description of Indian agriculture, Terry wrote, "In the northernmost part of the empire they have...good roots as carrot, potatoes and others like them...are grown". Although Terry was very likely referring to potato, he might have meant

the sweetpotato, which was by then being cultivated on a very minor scale in India.

British colonial governor Warren Hastings promoted (true) potato cultivation during his term, from 1772 to 1785, and by the late eighteenth to early nineteenth century, potatoes were sufficiently established in the hills and plains of India that varieties had acquired local names, such as: Phulwa ("flowering in the plains"), Gola ("round potatoes"), and Satha. However, the potato remained a garden vegetable of minor scale, often grown at higher altitudes by British colonizers as a summer crop.

Varieties were introduced via Europe, most of which proved unsuitable to Indian conditions on the hotter plains, since they were adapted to cultivation during the long summer days of Europe, not the shorter winter days of southern Asia best suited to larger scale potato production.

Initial attempts to establish the potato in India were also challenged by storage during hot summers and fast degeneration of seed tubers, probably due to viral infection.

A few varieties survived at higher altitudes, such as Magnum Bonum, Royal Kidney, Great Scot, Craig's Defiance, and Up-to-Date, the last of which was subsequently introduced from India east to Burma, where it remains a prominent variety.

The Central Potato Research Institute (CPRI) was founded in 1949, approximately two years after national independence, to develop potato varieties and technologies appropriate to Indian conditions.

By 1950, 32 cultivated varieties had been identified, 16 each of relatively recent European import and more locally adapted so-called Desi varieties. By 2002, CPRI had released an additional 35 varieties, contributing to an enormous expansion of potato cultivation and productivity.

Over the past few decades, potato has become the fastest growing staple crop in India. National production trends over forty years are summarised in the table below.

Table 1. India: Potato Production Trends

	1961-63	2001-03	Comparison
Total Area Cultivated ('000 Hectares)	383	1,281	334%
Per Capita Area Cultivated (Square Meters)	8.3	12.2	148%
Average Yields (Tons per Hectare)	7.3	18.7	256%
Total Production ('000 Tons)	2,844	23,980	843%
Per Capita Production (Kilograms)	6.1	22.8	372%

GEOGRAPHY AND PRODUCTION ZONES

Physical Geography

India's total landmass of nearly three million square kilometers can be divided into three main regions:

- The Indo-Gangetic Plain;
- The Himalayas to the north;
- The Southern Peninsula.

The Indo-Gangetic Plain, the site of most potato cultivation, is a great alluvial crescent stretching from the Indus River system and the Punjab Plain (traversing western India and Pakistan) eastward across the Haryana Plain to the delta of the Ganga (or Ganges) in Bangladesh, where it is called the Padma. The plain is characterised by a fairly consistent topography, differentiated slightly by floodplain bluffs and other features of alluvial erosion.

Climate and Precipitation

Although India's climate is highly variable over the enormous range of the country, there are distinct seasons:

- A relatively dry, cool winter from December through February;

— The hot, dry summer from March through May;
— The southwest monsoon from June through September when the predominating southwest maritime winds bring rains to most of the country;
— The northeast, or retreating, monsoon of October and November.

A failure of the monsoons can occur due to variations of the El Nino Southern Oscillation (ENSO), causing severe long-term and extensive drought, as happened several times from 1876 to 1902. Estimates of famine mortality in India during that period range from over twelve million to nearly thirty million people.

Distribution of Potato Production

Potatoes are widely grown in India, but the greatest concentration by far occurs along the Indo-Gangetic Plain to the northeast, where most potatoes are grown during the short winter days from October to March. Potato cultivation in the south is generally limited by an excessively hot climate.

Areas of high relative concentration of potato cultivation include:

— The states of Uttar Pradesh, West Bengal, and Bihar accounting for more than 75 percent of area under cultivation and about 80 percent of total production;
— Punjab state to the west, adjacent to Pakistan, where roughly seven percent of area and production occur;
— Higher altitude areas to the north, where a summer crop accounts for about five percent of total production;
— The southern peninsula, where approximately six percent of the total crop is grown on relatively high altitude plateaus, such as the Nilgiri and Palini hills of Tamil Nadu, under alternating rain-fed and irrigated conditions throughout the year.

Production System

Land Use

As of 1987, field crops were planted on about 45 percent of the total landmass of India, or roughly 135 million hectares. Of this land, almost 37 million hectares were double cropped, making the gross sown area equivalent to almost 172 million hectares. Discounting land which is developed for nonagricultural purposes, suitable only for pasture or otherwise unsuited to agricultural, potential expansion in crop production has to be derived almost entirely from lands already in some kind of agricultural use, a situation which has only intensified further in the intervening years.

Under these circumstances, the system depends on a greater diversity of crops which are productive relative to land and time and which complement the cropping calendars of major staples (wheat, rice, and maize), all factors which favour the rapid expansion of the potato crop.

India remains mostly a land of small farms cultivated by family labour. As of 1980, about half of all agricultural holdings were of less than one hectare, while large holdings (ten or more hectares) accounted for only four percent of all holdings.

Legislation pertaining to land tenure and the establishment of fair rents for tenants lies within the legal realm of states, not the federal government, hence is variable across the nation.

As a result of diminishing farm holdings and underlying population growth, the number of landless or near-landless families has been growing since independence both in absolute numbers and proporti-onately to the population. Without their own viable land holdings, many rural Indians are dependent on seasonal work at the peak planting and harvesting seasons. In 1981 there were nearly 200 million rural workers, of whom over 55 million depended primarily on occasional farm work for a livelihood. By the early 1990s, the rural work force had grown to 242 million, of whom nearly

a third (over 73 million people) were classified as agricultural labourers.

Cropping Calendar

Very generally, there are two major cropping seasons in India:

- — Kharif, during the south-west monsoon (June/July through September/October), when agricultural production takes place both in rainfed areas and irrigated conditions;
- — Rabi, during the winter, when agricultural activities take place only in the irrigated areas.

Climatic variations across India determine not only the regional distribution of the crop, but also to a large extent the seasonal cropping patterns. The optimal temperature range for potato is from 15° to 25°C (59° to 77°F). However, night temperature is critically important to tuber development and should not exceed 20°C (68°F), a factor which limits the season in many areas in the south, as frost can limit the season in the higher altitude regions of the north.

Across the Indo-Gangetic Plain, generally one crop is grown during winter, planted from late October to early November and harvested January through February. In the eastern areas, it is sometimes possible to take advantage of the shorter and milder winter by planting two potato crops in succession, the first roughly from September to November-December, followed by another crop from November to March.

The smaller high-altitude potato crops are typically planted in May or June, and harvested in September or October. Where mid-altitudes allow for an earlier spring crop, planting can be undertaken in February or March for a harvest in July or August, when prices are generally very high.

The autumn crop, planted around mid-October and harvested February or March, is important to the Punjab area to the west, adjacent to Pakistan.

Cropping Patterns

Land preparation

Prior to planting, land should be of a fine tilth for easy placement of seed tubers and to allow roots to penetrate the soil. Depending on soil type and crop history, this can normally be achieved by one plowing and one harrowing, followed by planking. In lighter soils, farmers do not always have to plow prior to planting, but if the potato crop is following paddy rice or jute, more preparation may be necessary. However, studies from Gwlior, in the north central area of the plains, have shown that minimal tilling (one harrowing followed by one planking) did not diminish the subsequent potato yield which followed a green manure crop of sunhemp plowed under in the previous kharif season.

Water management

Potatoes planted during the spring and summer in the mountains and plateaus to the north and northeast are usually grown under rainfed conditions, but the predominant winter rabi crop cultivated across the Indo-Gangetic Plains is grown with irrigation. Furrow irrigation is the most common method, based on long furrows where mechanical cultivation is practiced. Shorter and more tightly spaced furrows are more common under manual operation.

Sprinkler systems are sometimes used where conditions do not favour furrow irrigation, e.g. undulating topography or very sandy soils. Sprinklers also offer the advantage of more even water distribution and generally result in higher yields, though the cost of a sprinkler system is beyond the means of most farmers. Drip irrigation has been introduced, with impressive results in terms of enhanced efficiency of irrigation water. About fifty percent less water use has been reported, with twenty to thirty percent higher yields relative to furrow irrigation. However, drip irrigation is also an expensive option for the time being. Mulches have been proven to conserve water in experimental treatments.

Soil amendments and intercropping

Green manuring is recommended for potato, especially in the northern plains, though the actual extent of the practice is not reported. Commercial fertilizers are widely used, although typical applications rates and practices are not reported.

The most appropriate use of fertilizer varies by locale, but the general range of recommended applications for the northern plains area by kilograms per hectare (kg/ha) is: Nitrogen 180-240; Phosphorous 80-100; Potassium 100-150. Effective use of fertilizer is sometimes cited as a reason for intercropping. As potato roots are shallow, fertilizer dissolved in water can leach below the potato root zone, but might subsequently be recovered by a deeper rooting crop such as wheat.

In the western and central Indo-Gangetic plains, potato is a rabi crop rarely grown as a sequential monocrop. The most common rotation is maize-potato-wheat, though in some areas a maize-potato or paddy rice-potato rotation is practiced. A potato-sugarcane sequence is becoming popular partly due to the compatibility of the planting and harvesting, sugarcane closely following the potato harvest in spring.

In the eastern Gangetic plains, rice is more commonly part of a rotation, as is jute (grown for fiber) in West Bengal. In Bihar and Jharkhand, potato is popular as a rotation crop since it makes effective use of small land holdings and readily available family labours.

Crop esablishment

Since seed tubers are a major factor of production costs, farmers have to weigh the relative benefits of larger tubers, which produce faster initial growth and higher potential yields, but at higher cost. Generally recommended practice (again, not necessarily typical practice) is to plant tubers of 30 to 40 grams at a spacing of around 60 by 20 centimeters. Since the cropping season of the Gangetic plains is short, a common practice is to pre-sprout seed tubers prior to planting. Tubers are removed from cold storage and kept for ten to fifteen days

in diffused light to ensure that sprouts will be short, thick, and green.

Harvesting

Ideally, haulms are cut about ten days prior to harvest to promote maturation of the potato skin and therefore minimize bruising and damage during harvest. This is not always possible when a crop is to be sold immediately after harvest and is not expected to be in storage for an appreciable time. Both mechanical and manual harvesting techniques are used.

Control of Potato Diseases and Pests

This list is not complete, but includes several diseases and pests known to be serious constraints to potato cultivation in India.

— Late Blight is caused by a fungus-like oomycete, Phytophthora infestans, which is a specialised pathogen of potato and, to a lesser extent, tomato.

— Early Blight is a fungal infection caused by Alternari solani, the disease is manifest by circular to oval brown spots on older and lower leaves, occurring before tuber initiation but continuing to develop until the death of the affected plant. Mature lesions often exhibit a "bull's eye" appearance.

— Cercospora Leaf Spot, caused by Cercospora solani tuberosi, affects both foliage and stems. Spots, reaching two to five millimeters in diameter, are brown with a whitish center. The disease appears to be particular to India, first observed at Patna in Bihar state in 1951, but to date little information has been reported.

— Powdery Scab, caused by Spongaspora subterranea sp. subterranea, has been reported as a fairly widespread disease in India.

— Black Scurf is a fungus, Rhizoctonia solani, that attacks tubers, underground stems, and stolons of potato plants, especially in cool, damp soils. Although generally described as more of a cosmetic problem due

to black irregular encrustations of fungal sclerotia, it has been described as a major concern in India, attacking and killing young sprouts on tubers before emergence. Seed tubers can be treated with fungicides, but the disease can also be controlled via crop rotations with cereals and legumes.

— Dry Rot is a widely occurring problem of potatoes in storage, caused by infection of fungal Fusarium spp. Symptoms are generally visible about a month into storage, first as small brown lesions which can cause internal cavities and mummified rot. It can be avoided via careful harvesting and post-harvest handling, and can also be treated with fungicides and biocontrol agents.

— Viral Diseases. At least twelve viral diseases are known to infect the potato crop in India, causing progressive degeneration of seed stocks. The most important include: PVX, PVY, PVS, PVA and Potato Leaf Roll Virus (PLRV).

CPRI is routinely using micropropagation techniques, such as meristem tissue culture, to eliminate viral infections in "basic seed" tubers from which certified seed tubers are produced.

Prior to release, tubers are screened for the presence of viral infection via techniques such as ELISA. Of the varieties released by CPRI, the widely grown Kufri Jyoti, Kufri Bahar, and Kufri Sindhuri seem to exhibit a slower rate of degeneration than others. Since viral infections are often transmitted via aphids, CPRI has surveyed a number of locations for aphid population. Where and when population are minimal, seed potatoes can be grown with minor precautions, as in some locations of the northern plains from October to early January.

Aphid population dynamics have been modelled, relying primarily on temperature and humidity, to enable predictions of population changes one or two weeks in advance. This work has made possible the "Seed Plot Technique," noted

below under "Varieties and Seed Systems/ The Formal System."

Bacterial Wilt (Ralstonia solanacearum) is the most destructive bacterial disease of potato and several other plants of economic importance, known in India as ghera, uktha, bangle blight, bangdi, or parya. Losses attributable to bacterial wilt can be as high as 75 percent and have caused potato to be temporarily abandoned in some areas, such as Bihar.

Potato Cyst Nematodes have been reported as serious constraints to potato cultivation in the Niligiri Hills of Tamil Nadu. This is not a major potato producing area, but losses have been reported as high as 80 percent. Research has indicated that cyst nematode population can be reduced with positive net financial returns via the integration of a strain of rhizobacterium, Pseudomonas fluorescens, as well as the use of neem cake as a soil amendment and/or intercrops of mustard.

Potato Varieties and Seed Systems

Potato varieties in India have been derived from two subspecies, Solanum tuberosum subspecies tuberosum and subspecies andigena. As of independence, most varieties were of recent European import, and few were suitable to the short-day conditions of the Indo-Gangetic Plain. CPRI has concentrated its efforts on the development of varieties which are short-day adapted, heat and drought tolerant, resistant to degenerative viruses, and relatively storable during hot Indian summers. A few varieties have been cited as generally popular, among them Phulwa, Darjeeling Red Round, and Gola, and Kufri Jyoti, Kufri Bahar, and Kufri Sindhuri.

In Meghalaya State in the northeast, and possibly in other relatively high-altitude areas, the older established and more locally adapted Desi varieties retain more popularity, partly due to the slower rates of degeneration and shorter dormancy of some varieties. Slow degeneration is a more important factor where the higher-altitude environment allows for the possibility to retain seed stocks over several generations. In

some areas, shorter dormancy can allow farmers to plant both a spring and autumn crop. Among Desi varieties still being grown on a small scale by some farmers in Meghalaya are Lah Saw, Lah Syntiew, Lah Lupon, and Lah Taret.

Varieties have been selected for consumption in fresh form, but a more recent objective of varietal development has been selection for processing, i.e. varieties with high dry matter content and low levels of reducing sugars. The first two "processing varieties," Kufri Chipsona-1 and Kufri Chipsona-2, were released in 1998.

"Informal" Seed System

A "seed system" is broadly defined as "an interrelated set of components including breeding, management, replacement, and distribution of seed". Most tuber seed used in India has been produced and distributed by farmers themselves (the "informal system"), but individual farm households are not usually self-sufficient for a long period of time. Farmers need seed tubers from an external source at least occasionally for several reasons, such as to replace seed lost to disease or pest infestation, or to acquire new varieties.

Since the ecological diversity of India allows for several potato crops grown according to different calendars, farmers have traditionally depended for seed tubers on harvests from other areas made a few months prior to their own planting. Some areas of potato production, such as West Bengal to the east and Karnataka to the south, are not suitable for the production of seed tubers due to high population of aphids (the vectors of potato virus) during the cropping season. Potato crops grown at colder, higher altitudes have been important in the "informal" system, due to their complementary cropping calendars and the slower degeneration of tuber seed at lower rates of viral infection.

Seed is likely to be a high cost of production to the extent that it is imported across greater distances. Farmers in West Bengal and Karnataka depend on seed imported from the north, some transported as far as 2,000 kilometers. As seed degenerates at a fairly rapid rate in these areas, farmers have

to replace their seed with newly imported stock, generally every two years in West Bengal and every year in Karnataka. Unfortunately, this movement of tuber seed can spread disease.

In Meghalaya, a mountainous state in the northeast, nearly 98 percent of all seed used to establish new crops is provided by local production, in many cases by farmers who retain some of their own tubers for seed. Most farmers in Meghalaya have used the same seed stock for ten generations, and some have not replaced their seed for over twenty years.

When degeneration becomes so severe that farmers have to acquire new stocks, they often spend very precious cash to buy seed from traders or other farmers. CPRI is working to introduce new seed production techniques, while the local Department of Horticulture provides farmers with subsidised seed, but their efforts are challenged by very high transportation costs from distant areas of northern India, such as Himachal Pradesh.

"Formal" Seed System

The "formal" system refers to seed tubers produced and distributed by state-sponsored institutions (possibly with some involvement of the private sector and/or non-government organisations). Seed from the formal sector has generally been subject to an inspection process intended to assure that the seed is of the variety claimed, with little or no incidence of disease or pest infestation, and otherwise viable. "Basic" or "breeder" seed is multiplied into "foundation" and "certified" seed, available for distribution to farmers, although the precise definitions of these terms are variable by country and region.

The formal seed system, in India and other countries, also serves to introduce newly developed or acquired varieties. CPRI has been very actively involved in the selection and release of new varieties, 35 to date since 1949. Most potatoes planted in the Indo-Gangetic Plain are of varieties released by CPRI, estimated at 93 percent in 1999 (WPC). In that year, CPRI produced approximately 2,600 tons of basic seed. CPRI

distributes basic seed via the departments of agriculture/ horticulture of each state, which are in turn responsible for multiplication and distribution of certified seed. In practice, the quality and availability of certified seed is therefore quite variable by state.

In some cases, local multiplication and distribution is accomplished with the active cooperation of local farmers. For example, in Meghalaya State, the Department of Horticulture supplies foundation seed to selected farmers who multiply and resell it to the department for subsequent distribution to other farmers.

Depending on local needs, farmers might continue multiplying foundation seed into subsequent generations of seed which can still qualify as certified, e.g. certified 1/2/3. A national study undertaken in 1999 estimated that fewer than 25 percent of farmers were able to obtain healthy seed as and when needed to replace their debilitated stocks (WPC), suggesting that the seed system remains a constraint to further expansion of potato production in India, though not to the same extent as in many other tropical countries where potato is an important crop.

As a strategy to address this constraint, the "Seed Plot Technique" has been developed as a means to provide seed free of viral infection by harvesting tubers intended for seed prior to the development of aphid population, building on the work noted above in the "Production Systems and Constraints" section under "Viral Infections". Larger quantities of higher quality seed are now being produced on the plains of Punjab and Haryana in the north via this technique. True Potato Seed (TPS), an alternative to clonal tuber reproduction, has been investigated by CIP for potential applications in several countries, although India is one of a few where TPS has been considered a potentially profitable alternative to reproduction strictly via the usual clonal propagation techniques.

Consumption and Storage

By estimated 2003 data, India's population of 1,065,462,000

consumed 25 million tons of potatoes, or about 23.5 kilograms annual per capita consumption. This is high by world standards, though still much less than countries where potato is the main staple.

The high growth rate of potato production in India is likely to continue, primarily as a marketed cash crop more than as a rural staple. Although very little of the potato crop is processed on a larger industrial scale, homes and small businesses sell snack foods such as dehydrated and fried potato slices or shreds.

Potatoes have been adapted to Indian cuisine in several creative ways. A few dishes from the northwest include:

— *Aloo Ka Paratha:* prepared by pinrolling spiced and boiled potatoes in wheat dough and frying in butter. It mainly serves as an item on breakfast menus consumed with fresh yogurt or sweet butter milk;

— *Aloo ki Tikki:* a traditional snack item consisting of a patty of boiled potato, finely chopped onions, garlic and ginger previously marinated in a mélange of spices, and deep fried until golden brown in white butter. It is served hot with a sauce of tamarind and red pepper;

— *Dum Aloo*: a curry prepared with boiled potatoes mildly fried and cooked with sour curds, red tomatoes and served with a thick spicy gravy, garnished with finely chopped green coriander.

Storage

A major challenge to potato cultivation in India has always been that, unlike temperate regions where the potato harvest is followed by a cold winter, the harvest in February and March is followed by the hot summer months. Refrigeration is necessary for potatoes which have to be stored until the following October.

Cold storage now takes place on a large scale, with over 3,400 facilities providing capacity to store over ten million tons (roughly 40 percent of total production). Approximately 60 percent of potatoes in cold storage are ware potatoes, intended

for consumption, with the other 40 percent intended for seed. Storage for seed is usually at colder temperatures, 2° - 4°C (38° - 43°F), in order to prevent premature sprout growth. However, such cold temperatures induce high levels of sugar accumulation, rendering potatoes less suitable for consumption.

Storage at a higher temperature range, around 10°C (50°F), is therefore better suited to ware potatoes. To inhibit sprout growth, a suppressant is often applied, most commonly CIPC (isopropyl N (3-chlorophenyl) carbamate), which checks growth by inhibiting cell division. CIPC has been in use since 1998. Cold storage facilities are usually commercial operations subject to some government regulation regarding rental rates and quality control, again variable by state.

Local governments also provide subsidised loans for part of a cold storage investment. Farmers usually store their own produce in bags each weighing fifty kilograms. The local price of electricity is a key factor in rental rates and the viability of facilities; high electric bills have been cited as a reason for closure of some stores. Demand for cold storage facilities is high, and stores are usually filled to capacity during the peak season.

Marketing of Potatoes

Marketing of potatoes is essentially a private enterprise function, with the role of government agencies generally limited to providing information of supplies and prices, and some regulation, e.g. the promotion of standardised grading. However, as with other aspects of government administration, variations occur across state and local urisdictions.

Unlike several other commodities, government policy does not provide price supports for the potato crop. Nor has the government enforced quality standards, which some claim has hindered India's potential to develop a potato export industry.

The strong seasonality of potato production combined with limited storage capacity often induces strong fluctuations in seasonal availability, hence price swings. Following the end

of the main harvest in February, prices begin a rapid increase in April, and have often nearly doubled by July-August. However, interannual fluctuations can also be strong, characterised by "potato gluts" and resulting price crashes, followed by reduced plantings and supplies the following year.

Potato Production in Kenya

At the commercial level, potatoes are mainly consumed as chips served in restaurants and take-away facilities in major towns in Kenya. The white varieties such as Nyayo and Tana are the most popular potatoes for chips, although a few takeaway facilities prefer the red varieties, which they claimed take less oil and give higher volume of chips than the white varieties.

The potato crop is grown mainly in the cool, high altitude areas with well distributed rainfall. The most suitable elevation is between 1,500 metres and 2,500 metres above sea level. The main potato growing areas are found in Central, Eastern and Rift Valley provinces. Central province produces more than 53 percent, while Eastern and Rift valley provinces produce a combined total of 44 percent.

In Eastern Province, the main growing region is the Meru district in the areas around the slopes of Mount Kenya. In Central province nearly all the districts produce some potatoes, with Nyandarua district, which lies along the Aberdare mountain range, being the largest and most diversified potato producing area.

In the Rift Valley, potatoes are grown in the Mau escarpment region in Dundori, Mau Narok and Molo, and in the western highlands of Kericho, Bomet and Uasin Gishu districts. The most dominant varieties grown in Meru is Kerr's Pink and Ngure, which has close similarity with Kerr's pink, but which has a higher resistance to blight.

In the Meru district potatoes are mainly produced in the Kibirichia sublocation and the Timau division located on the northern slopes of Mount Kenya. The main variety grown is

Kerr's pink commonly known as "Mew potato" in terminal markets, but some farmers are also planting the Ngure variety which has a faded pink colour. The latter is competing with Kerr's pink, because it is more resistant to diseases and requires less chemical inputs. Seed potato is both an input and output of potato production. Particularly in the case of red varieties, it is doubtful if the market can be sustained without adequate certified seed; although it does appear that the acute shortage reported in 1997 is no longer there, there is a general lack of awareness on the available sources of certified seed.

Planting for the long rains is done in October and November, followed by harvesting in January and February. The short rain crop is planted in March and April and harvested in July and August. A few farmers have installed irrigation systems using gravity by tapping water from the streams.

The irrigated crop is planted mainly in January and February and in June and July, but farmers with adequate land can plant at any other time of the year. The harvesting time for the irrigated crop is timed to coincide with low market supplies and better prices. Potatoes are planted as pure stands in ridges. The seed rate is 8-10 bags per acre. Di-Ammonium-Phosphate (DAP), a common fertiliser, is applied at the rate of 100 kg per acre. Farmers who keep livestock also apply farmyard manure before planting.

The potato growing farmers in Meru practice mixed farming, with potato being the main cash crop. In each season the average area planted with potato by each farmer ranges from 0.25 to 2.5 acres. All farmers plant at least two crops a year, but those with irrigation systems grow three crops a year. Rotation is done after 2 or 3 crops grown continuously on the same plot, but farmers with small farms are not able to rotate their potato crop regularly.

Farmers mostly affected by small land size are in the Karuri area in the upper parts of the Timau division. These farmers have been growing potatoes on the same plot of land, twice a year for more than ten years, and the yields have been

decreasing every year. The area is also too cold and very few other crops can be grown, which limits chances of rotation.

The Karuri farmers do not have title deeds for the land they occupy, and they are not willing to invest in irrigation systems until they are sure that they own the land. Family and hired labour is used for the various processes such as land preparation, planting, weeding, harvesting, sorting and packaging. Harvesting rates range from 20 KSh to 50 KSh per bag harvested in different areas within Meru district with the lower payments being made in the Timau region.

The main production problem affecting potato farmers in Meru district is lack of certified seed. This means that the same seed is used for many years and may be a constraint to increased potato production, as it encourages disease multiplication. A drop in yields over the past few years in all the areas visited was attributed to lack of quality seed, since the seed currently in use is sometimes infested with bacterial wilt.

Other important constraints are late blight and virus diseases. The high price of fertilisers and fungicides makes it impossible for some farmers to follow the recommended application rates, and this also affects the yields.

Harvesting

Harvesting is done when the crop is mature and the stalks are dry. This is done by sharpened sticks or fork jembes to minimise damage to the tubers. Potatoes are harvested all at once per plot. The average yield is 60-70 bags per acre of good sorted crop for sale, but a few intensive farmers may realise up to 140 bags per acre in a good year. The Karuri area in the Timau division reported yields as low as 20 hags per acre during had seasons. When market demand is high, irrigated potatoes may be harvested before they are fully mature.

On-farm Post-harvesting

After harvesting, the potatoes in Meru are sorted into 4 grades. Grade 1 consists of the trig potatoes for sale. Grade 2 potatoes

are egg-sized. They are kept as seed and excess quantities are sold. Grade 3 is sorted for home consumption and some may also be sold as seed, while grade 4 includes all the small potatoes which are consumed at home or fed to livestock. Potatoes which are damaged during harvesting are prepared for home consumption or fed to livestock. This is not considered as a loss by the producers.

According to the farmers, the losses are very low (about 5 %), and affect the potato crop which has been stored for more than three months. Harvesting is done when the weather is dry, hence no cleaning is required. Discussions with farmers in Nyandarua revealed that sorting and grading is minimal. Sorting includes selection of seed, as well as the removal of very small and damaged potatoes.

In the Meru district, farmers have developed their own storage structures. In some homes, the structures are made of cemented floors, wooden walls and corrugated iron rooves. Others have raised wooden floors and similar walls and rooves. In the upper and cooler Timau region, farmers store their potato crop and seed outside the house covered with dry grass. Under these conditions potatoes can be kept for up to 3 months, due to the cold and windy weather.

Adoption of the improved storage structures developed by the National Potato Research Centre in Tigoni, appeared to be low or non existent in some areas. In Kibirichia two farmers had constructed small improved stores, which looked like demonstration units for other farmers to copy. Since the stores were very small, the farmers continued to store the main crop in the old stores.

Lack of adoption of the improved structures may be explained by the fact that farmers had already developed their own stores before the introduction of the new ones, and are not willing to incur extra costs. Secondly, potato storage is limited to a maximum period of 2-3 months, while waiting for better prices.

The same practice affects the potato stored for home consumption. The stored quantity is just enough to last until

the next crop is ready. The seeds are stored in pits which are lined with dry leaves and covered with straw until they sprout.

In Nyandarua, potatoes are sold soon after harvesting and the farmers have not developed storage structures. As in Meru, the seed is stored in covered pits until it sprouts. In Nyandarua, potatoes are sold soon after harvesting and the farmers have not developed storage structures. As in Meru, the seed is stored in covered pits until it sprouts.

Kenya's development objectives for the agricultural sector include internal self-sufficiency in basic foods according to the comparative advantage in the production of crops. Increased production is needed to cope with population growth. Since a considerable part of the population growth takes place in urban areas, effective and efficient post-harvest systems are required.

Potential major constraints might place a high burden on economic and social development, thus restraining economic growth. Potato and sweet potato post-harvest systems are especially vulnerable to be hit by constraint factors because of their bulkiness and perishability. Potatoes store well, but sweet potato has a shelf-life of little more than one week after harvesting. There is a constant supply of potato and sweet potato throughout the year. The actors in the post-harvest system therefore are continuously receiving market signals down the chain.

There is a strong all-year round demand for potato in the domestic, restaurant and institutional sector. Although potato is usually associated with middle class consumption patterns in developing countries, in Kenya it is also an integral part of the diet of those living in the central area—mainly Kikuyu.

As average real incomes rise, it can be expected that the demand will become stronger. Sweet potato, on the other hand is seen as an inferior food. In the traditional cook-book it is considered limited, and more of a snack than a proper staple. Sweet potato carries a "stigma" of being a food of last resort, or famine food.

An improvement in marketing conditions for sweet potato has been found in one of the study sites. Whether the supply response was due to an increase in overall demand fostered by the increasing price of maize and bread—or due to the more efficient marketing system. A major factor of the future development of sweet potato as a cash crop is the uptake of urban consumers as a basic staple food. The following factors are likely to affect further demand:

— demographic changes due to population growth will lead to an intensification of production,

— increasing urbanisation, accompanied by stagnating incomes will lead to the substitution of maize by sweet potato, because sweet potato provides cheap calories,

— the uptake of urban processing for sweet potato flour and the incorporation of the flour in traditional and new snacks.

The question remains, who are the consumers? The urban poor, who can no longer afford the more expensive items, and will switch away from sweet potato if the price of maize and other expensive food should fall? Or, is it the population originating from the western part of the country and residing in the cities, for whom sweet potato is a familiar, if not regular part of the diet, and who are buying it because market liberalisation has brought more things onto the market?

Available historical data on prices and production of the major food crops in Kenya were brought together by the Policy Analysis Matrix team of Egerton University in order to determine the substitutive relationship between production and demand of food commodities in the long run.

As a consequence of maize being the major food crop, small variations in production appear dramatic compared to the production of other crops. It seems that potato and sweet potato production move in the opposite direction to that of maize production. It can be supposed that as maize prices fall, farmers substitute with short term root crops—potato and sweet potato, expanding the area of these short term crops to the detriment of maize production.

Similarly when maize prices rise, farmers plant more maize reducing the area of potato and sweet potato. It might be that cassava also substitutes, but because of the long lag period between planting and harvest it is not so apparent on the limited data.

In late 1994, the price of maize fell after a bumper maize harvest. Prices of potato and sweet potato, on the other hand went up, as production fell. This suggests, that although sweet potato may to some extent substitute maize, there is a fairly inelastic core demand for the product. The demand for potato is relatively inelastic because of the constant demand from the hotel and catering trade for chips, and so prices are much more supply driven.

Post-harvest Chains

Potato is a much more commercialised crop, both from a production and marketing point of view. The major part of the crop is grown for sale. In contrast, sweet potato has traditionally been grown on small plots for domestic consumption. Production for the market is a recent development. In Kisii it was started in 1992, and only a few areas have really benefited from the market development.

The main stimulus to production for the market was the presence of traders who were regularly going to the large urban markets. This coincided with poor prices and late payment for coffee crop—a major cash crop in the district. Farmers were therefore looking for alternative crops to raise cash from. Finally, in 1994, a bumper maize harvest in Kenya severely depressed maize prices, thus making sweet potato appear a more attractive cash crop.

Potato has a well developed market chain from producers to the wholesale market. The bulk of potato marketing is done on a regularly scheduled system. Agents or brokers in the villages act for traders by organising farmers to assemble their crop ready graded and packed for collection, to fulfil orders made by traders. The lorry then delivers the crop direct to Nairobi market. Some consumers, notably chip manufacturers, have developed direct links to some farms to

ensure a regular supply and an assured standard. Such contracts represent a very small proportion of the business.

For sweet potato a similar system has developed in Kisii in the most commercialised areas, except here crop is usually transhipped in Kisii town from a pick-up to an intercity lorry. This is because the traders are smaller operators and usually do not buy sufficient quantities to fill a lorry on their own. In the less developed system, agents buy from farmers and deliver by donkey to the tarmac, where traders buy prearranged volumes at prearranged times.

The trader then flags down a passing truck to carry to Ahero or Kisii for transhipment to intercity transports plying between the main towns. Potato farmers in Meru receive 69% of the wholesale price both in the local and the Nairobi market, for their product, indicating a high degree of competition and low freight rates. Sweet potato farmers receive 55% on the Nairobi market. The trader's margin for potato varies considerably, with the highest margins going to the local Meru trader. Those trading from Meru to Nairobi seems to make the thinnest margin. This reflects higher costs due to repacking, but also may reflect a higher price to the farmer due to the enforcement of a more uniform standard of bag.

Efficiency of the Post-harvest System

— *Effectiveness in Meeting Demand Patterns*: The effectiveness in bridging supply and demand for the crop with special regard to the geographical distance between production and consumption, to seasonality and product quality is good in both systems. Potato production has responded to strong urban demand and is a major traded commodity. Steady demand increases due to a shift in urban consumption patterns have led to a constant development of supply chains from the major production regions. In sweet potato marketing, restricted urban demand is guiding the development of the post-harvest system. Commercialisation has reached only part of its potential.

— *Food Security*: The population of Kenya increased from 22 million in 1988 to 29.2 million in 1996, an increase of 3.9 % per year. Over the same period the GNP per capita fell from US$ 410 to just US$ 260 per head—a real decline of more than 5 % per year. Production figures for the major crops in the 90s has been falling. The consumer price index (CPI) for food has risen faster than the general CPI. This is an alarming picture. By the year 2010, the population is expected to reach 44 million and by the year 2025, sixty-three million. Maize is Kenya's staple crop, providing three-quarters of the carbohydrates for Kenya's fast growing population. The demand for maize is relatively inelastic, and shortage of maize results in large price increases. Maize production has spread to the margins in Kenya, and production suffers badly at times of drought. Sweet potato and Cassava are relatively more drought resistant. Sweet potato can play an important role in food security strategy for Kenya:

  — It is considered more nutritious than cassava.

  — It is more drought resistant than maize.

  — It can improve the yield of maize in a crop rotation compared to continuous maize production. Further, continuous maize cropping on the same field leads to reduced yields. As noted by some farmers in Kisii rotating sweet potato crop with maize improves the maize yield. Increased production of sweet potato could therefore improve farmers incomes through higher yields of maize, as well as income from sweet potato.

  — Finally, sweet potato is a relatively short term crop, with a flexible time of harvest allowing a high degree of flexibility in a food security strategy.

— *Barriers to Entry*: Given the number of agents, traders, porters and brokers involved in the system, there appear to be no serious barriers to entry. Obviously those with capital can acquire larger consignments and

reduce their overheads. Those with vehicles running efficiently can increase their margins. No serious efforts have been made to set up co-operatives to compete with the private trader and recoup some of the trading margins for the farmer.

— *Transport Costs:* Transport charges are competitive on the intercity routes because the vehicles will generally get a return cargo. The rental for a one-way journey would be 12,000 KSh producing an "operating profit" (not including fixed costs) of 27 %.

— *Loading Costs:* Loading is done by organised groups of workers in all market locations, and payment is usually made to the organiser. The rates therefore tend to be standard. Loading is usually 15 to 20 KSh per bag, and unloading costs 10 KSh. This might be one reason why traders prefer to carry large bags. In fact, smaller bags can be loaded at a lower price, after negotiation.

— *Transaction Costs:* A post-harvest system is allocatively efficient if it is able to minimise the transaction costs from production to final destination. No exploitative tendency could be found. Calculations show that in the potato post-harvest system more added-value is gained by actors belonging to the rural areas compared to urban actors, than in the sweet potato system. In both marketing systems, there are many agents and traders serving the farmer. It appears that prices to the farmer are uniform within a village, and agents/traders compete on service. Traders pay agents fixed commissions, but may top up in times of shortage.

— *Social Efficiency:* The post-harvest system of both crops involves many actors and creates "semi" and "casual" employment for large parts of the poorer sections of the population. It also provides a training ground in entrepeneurship to persons who may want to start another enterprise or expand up the marketing ladder. Little initial capital and skills are required to take part in the system.

Labour Requirement

The post-harvest system of root and tuber crops contributes significantly to employment in rural and urban areas. A comparison of pre- and post-harvest labour input shows that more labour is required in the post-harvest chain than in production. The model calculation based on crop budget data from Meru and survey data from marketing to Nairobi yields a ratio of 1.9 of post- to pre-harvest labour input.

— An estimated 50 % of the total amount is marketed by brokers/agents.

— An estimated 50 % of the total amount is processed to chips and crisps.

Assuming perfect competition in labour markets and nearly equal qualification, returns to labour should be the same in production and post-harvest. The value-added in the potato post-harvest chain is slightly higher than farmers' profit per unit.

Table 2. Value-added and return to labour in pre- and post-harvest operations of potato

Production:	
Profit per acre:	KSh 17,100
Profit per bag:	KSh 285
Profit per manday (return to farm family labour and land)	KSh 285
Post-harvest:	
Value-added per bag in the chain (between farm gate and retailing to consumers)	KSh 1600
Value-added per manday (3.24 mandays/bag)	KSh 305

Returns to labour are 285 KSh/bag in production (only for farm family labour) and 305 KSh/bag in post-harvest operation. However, returns on land are not accounted for in the figure of production costs.

Packaging

Packaging of the potato crop at the farm level is done in flat hags provided by the traders. The decision to use flat instead of the extended hags was made by the District Agricultural Committee (DAC) chaired by the District Commissioner, with the District Agricultural Officer, as the secretary.

Other members of DAC are Agricultural Extension Officers from each of the divisions, farmers representatives and County Council representatives. From December 1995, the DAC in Meru decided to introduce the flat bag policy. Since then all farmers are expected to sell their potato crop in flat bags.

Although this decision is not a formal law, it is enforced by communities, chiefs and youth groups. Currently all potatoes in Meru are packaged in the flat bag, and both the farmers and the traders seem to have accepted this practice. When the flat bag was initially introduced, prices were reported to have dropped briefly for a few months but then returned to normal levels. The packaging is normally done by the farmer and other family members or by hired labour.

The farmer decides how much to sell at any given time, and the price is agreed between the farmer and the agents/ brokers on behalf of the trader. The agents sew the bags and load them into the trucks and they are paid an average of 50 KSh per bag.

In the dry season, potatoes are collected from the farms by the trader using trucks or pickups. In case of wet conditions, farmers use bullock-drawn carts to take the produce to the nearest market centres. When the traders leave Meru district boundaries, they re-package the bags and extend them to the weight of 120-130 kg, which is the weight accepted in Nairobi and other markets. The main market for potatoes from Meru are Nairobi, Mombasa and other districts neighbouring Meru such as Nyambene and Tharaka-Nthi.

The crop seems to be in demand throughout the year, but the prices vary according to the season. Potatoes in

Nyandarua district are still being packed in extended hags. The extension normally takes another half bag. Packaging may be done at home during the dry weather, but when wet, farmers transport the crop to the nearest murram road and package there. The brokers (agents) are responsible for sealing the bags.

Rural Consumption

In the homes especially within the producing areas, potatoes are consumed daily for both midday and evening meals. They are made into stews and mashes which form the staple foods for most people in these areas.

In the local dishes potatoes are mixed with vegetables such as carrots and cabbages and meats and made into stews which are eaten with ugali, chapati or rice. They are also used in traditional dishes in which they are mashed together with maize and bean, or peas and other pulses into which some varieties of green vegetables may be added. Experience shows that this mode of consuming potatoes is duplicated in urban homes of families whose origins are from potato growing areas.

Marketing System

The marketing system is influenced by the nature of the crop. Potato is a high yielding, high bulk, and relatively low value commodity. It takes 3 to 4 months to grow. After harvesting the crop may be stored for relatively long periods, loose, in a proper store, for periods of up to 2 months without loss.

The relatively high bulk, means that transport efficiency a key to general market efficiency. In the Meru district, potato is the main cash crop for farmers, and the bulk of potato grown is for sale. Farmers mostly sell through agents or brokers who work directly to fulfil orders for traders.

A few farmers close to the market towns might bring crop into the market and sell directly. In Nyandarua (Kinangop), potato is a major cash crop, and the marketing is very similar to Meru.

Market Places

Farmers at present are selling in small quantities of about 5 to 10 bags, in the expectation of a price increase. Most of the sales are arranged by a local broker who is acting on behalf of a trader from Nairobi and other neighbouring districts. There are also traders from the local markets and Meru market who come directly to the farmer and buy.

The Meru retail market operates three days a week: Monday, Wednesday and Friday, with Monday and Friday as the main trading days. The throughput of potato on any market day is between 1,000 and 2,000 bags.

The traders in this market buy direct from the farmers in the growing areas (Kibirichia, Nari and Timau) and transport produce by pickups, small lorries or donkey/bullock carts. Much of the produce in this market is sold to traders coming from smaller towns in the district (Nkubu, Maua, Isiolo, Chuka, Chogoria and Embu). Sometimes farmers bring up to 10 bags for sale in this market, but they usually have their own donkey or bullock cart.

Traders buying crop from the farms in Meru stop at Nanyuki to rebag the potatos into extended hags. The traders have been buying potatoes in the flat bag since 1995 when the District Agricultural Committee enforced a uniform system on the whole district to protect the farmer from being exploited. However, the Nairobi market "requires" an extended hag.

The bags are therefore off-loaded in Nanyuki, then opened and a "sisal net" extension is woven on the top of the hag. The bag is topped up with potatoes to weigh about 130 kg, an increase in weight of about 30 kg. From 100 "flat bags" traders get 72 extended bags.

The main destination market is the wholesale market of Nairobi. Nairobi's major wholesale market is Mincing Lane or Wakulima Market, built in the 60s, when Nairobi's population was much lower. It is now far too small, and much of the trading actually goes on in the streets surrounding it. Lorries arrive at night from 11.00 p.m. onwards. The drivers hand

over the lorries and freight to brokers who do the wholesale marketing and then leave the market.

The market opens at 4.00 a.m. and some lorries enter to sell there. Others remain on the street parked with their tailgate to the street, and brokers sell by the bag from the back of the lorry.

Other major markets for the Meru potato include neighbouring districts of Nymbene and Tharaka-Nthi where potatoes are not grown, and where farmers have adequate cash from Miraa sales. Nairobi is also a transhipment point for other centres, especially the coastal town of Mombasa as well as a final market.

Some of the produce coming into the market is taken by hawkers who spread out a mat and sell by the tin to consumers and minor retailers.

Most vegetables in Kenya are sold in markets rather than supermarkets or proper shops. These markets exist in all the suburban areas, and vary in terms of their physical state and volume of trade.

Margins in the Chain

Table 3 shows the selling prices at different stages in the marketing chain.

— *Farmer Level*: The February price per bag in Meru was 900 KSh for a 100 kg bag. In Kinangop, the farmgate price in early March was 1,000 KSh, but edging up to 1,200 KSh for the Tana variety. Currently prices are abnormally high, due to the poor rains last year, and the shortage of maize on the market. The price of potato is usually between 400 and 600 KSh per hag, depending on the accessibility of the farms. During the rains, when the accessibility is poor, the difference between the farmgate price and the local market price (Mew market) can be as much as 50%.

Table 3. Margins for the Meru pink potato variety and the Tana from Kinangop (KSh)

Level	Meru	Kinangop
Weight per bag:	*100 kg*	*130 kg*
Farm gate	900	1,200
Meru wholesale	1,300	
Nairobi wholesale	1,385	1,800
Nairobi retail	2,500	3,000

— *Village Level Broker:* A village broker receives 50 KSh per bag. This includes his contribution to loading the lorry. Organising 20 to 30 hags in two or three days a broker can earn between 1,000 and 1,500 KSh. Brokers can organise up to 2 collections per week.

— *Meru Market Trader:* Sells produce for 1,300 KSh per bag, realising a profit of 250 to 300 KSh per bag, each market day. If trading 20 bags per trip, he can earn 5,000 KSh per trip (one day collect, one day sell). A trader can usually manage 3 trips per week.

— *Trader Selling in Nairobi*: Potatoes being bought in the Meru district at 1,238 KSh per hag, were selling in Nairobi for 1,800 KSh yielding a profit margin of 200 KSh. Potatoes purchased from Kinangop at 1,200 KSh (Tana) were selling at 1,800 KSh per bag, giving a profit margin of about 300 KSh. Margins can be higher by 20 to 50 KSh, if traders avoid paying the Nairobi market cess, and if they are able to get hack the hag from customers for free or a small fee, which some traders were able to do.

— *Margins for Small Town Traders*: Margins were calculated for traders selling in Chuka and Machakos markets, on the basis of interviews held. The margins here were found to be low on the wholesale side, but high on the retail.

— *Agents*: Buy at 1,400 to 1,500 KSh per bag and sell at 1,800 to 2,000 KSh to their customers after grading and

repacking into smaller bags. The rejects from sorting are estimated at 12-15 kg per extended bag, which are then sold to traders for 100 KSh. Overall profit is 300 to 400 KSh per bag depending on transport costs from the market to the final customer.

Functioning of Market

It is very difficult to calculate the margins for traders due to the inexactness of the hag measurement used. Although fixed prices were always quoted, there is some lee-way in the selling on prices. When wholesalers sell by the tin, they frequently add potatoes from another tin, to give the impression you get more for your money.

— *Weights and Measures*: Traders were reluctant to tell their true margins, and their estimate of 7 to 8 tins per bag, yields almost zero margin. It appears that a bag can fill 9 tins or possibly even 10 tins Efforts by Meru District Agriculture Committee to standardise the bag size may have improved their margin and lowered that of the middlemen.

— *Efficiency*: The market appears to operate extremely efficiently. There were no complaints that farmers could not sell their product. Traders also did not complain of problems disposing of their crop. In times of glut, prices come down to clear the market.

It seems that there is a lot of information exchange between brokers and traders. Changes in demand or supply are quickly communicated, because of the regular system of collection from farms.

Processing of Potatoes

The industrial processing of potatoes is restricted to the production of snack type foods, such as crisps and other types of snacks specifically for Asian consumers, such as chevra. These industries are diversifying their product range to include roasted peanuts and spices. There are more than

twenty crisps enterprises located in Nairobi. However, the number could be higher since some processors are operating in private homes and may not be registered.

The larger potato crisp makers use an average of one metric ton of potatoes per day. The production depends on the amount ordered by their customers, who are mainly the major supermarkets, hotels and clubs in Nairobi and other major towns. The total demand probably does not exceed 5 tons per day.

The dry matter content of potatoes is important for potato processing. Tubers with high dry matter content are preferred because they are mealy and do not consume a lot of oil during frying. The red potato varieties such as Kerr's pink, Golof and Ngure are preferred by crisps processors because they meet these qualities. Secondly, their shallow eyes make it easier to peel. The major source of the potatoes used for processing are Meru, Bomet and Nanyuki.

Some of the processors purchase their potatoes from Wakulima market in Nairobi while others use agents to supply what they need. One of the processors interviewed buys directly from a large scale farmer, at current market prices. The farmer delivers the potatoes to the plant in flat bags, and the processor is happy to pay at current prices since the produce is well sorted out before delivery.

The processors who were interviewed indicated that the market was improving, in spite of the fact that the number of processors was increasing. This shows that the market is growing. However, the future growth of the crisp industry is threatened by the importation of products with better packaging, which is competing with local products. This competition is healthy especially for the consumers because it is forcing the processors to improve on the quality of their packages.

Operations in the Potato and Sweet Potato Post-harvest Systems

The operations in the potato and sweet potato post-harvest

chains are almost solely executed by private sector agents. On the national level these two crops have not yet been considered as priority crops in the government research and extension system compared to other crops such as maize and wheat. Only to a limited extent public institutions have supported research and extension on potatoes and sweet potatoes. However, there are recent examples where local public entities have intervened in the marketing system.

Public research on both crops has been carried out by the Kenyan Agricultural Research Institute (KARI), partly in cooperation with the International Potato Center (CIP). Research on post-harvest problems focused on technical issues such as flour processing and post-harvest protection. However, links to extension are weak.

Recently, the potato sub-sector faced the break-down of certified seed production. KARI and other para-statal organisations were supposed to supply certified seed to farmers, but in fact did not effectively reach them. The private sector is not in the position to fill this gap. This leads to a deterioration of the seed quality. In Meru, farmers complained about lacking high-quality seed potatoes thus causing a breakdown of disease resistance and low yields. KARI-Tigoni is just starting a NGO-sponsored project which aims at informal seed distribution among farmers at the local level.

Extension activities in sweet potato production were rare mainly because sweet potato for a long time has been considered a subsistence crop, which is hardly marketable. For this reason an extensionist recommended to farmers not to spend money on inputs for sweet potato production, but to dedicate them to marketable crops. The very few efforts made in sweet potato extension addressed on-farm processing in rural women groups. Those are still working, but to date the quantities processed are negligible.

In 1995 the Meru District Agricultural Committee (DAC) decided to intervene in potato trade. Until that date extended hags were used in potato marketing, which was unsatisfactory for most farmers because the size of the extension was difficult

to standardise. After numerous discussions within the DAC, it was decided to introduce flat bags in Meru district.

Remarkably the decision in favour of flat bags was not implemented through a formal law, but through the collaboration of various groups of the society. At the farm and in the villages the chiefs and the members of youth clubs made sure that only flat bags were used when packing at the farms, outside the villages the trucks were checked by officials. When extended bags were found they had to be repacked on the spot.

Table 4. Composition of the District Agricultural Committee in Meru District

Chairman: District Commissioner
Secretary: District Agricultural Officer (DAO)
- Division Agricultural Officers (one per division)
- Farmer representatives (one per division)
- County Council representatives
- Members of Parliament from the district (occasional)

Meru can be an example for other potato producing areas where the same problems are common. If more District Agricultural Committees support the flat hag standard would be enforced by public-private sector partnership on a national scale.

Stability of the system

There is a difference in the type and number of operations and actors in the PH system between a highly commercialised crop (potato) and a crop in the initial stage of commercialisation (sweet potato). The structure of the marketing of potato shows more direct links between the farmer and different market actors. From the farmers' point of view the market is therefore well developed. Co-operative marketing has not been established.

The diversity of acting agents and middlemen involved in potato marketing is considerably high. There is effective

competition among actors at the different levels. From the farmers point of view, there are usually several marketing options including storage at the farm. The system is therefore relatively robust to external shocks.

Although farmers were not so much aware of potato prices in distant urban markets, information about the price level at local markets is efficient. Farmers know that supply shortages in substitutive commodities like maize boost potato prices. The sweet potato post-harvest system is still in the stage of limited commercialisation. The post-harvest chain is mainly operating through a few bottlenecks which are the assembling and transhipment points in Kisii and Ahero.

Constraints in the post-harvest system

Potential constraints of the sub-sector have been collected by the survey team in a two-stage approach:

— an expert meeting during the preparatory phase with participation of representatives from MoALDM, KARI, universities, CIP, GTZ and NGOs

Constraints in the post-harvest system can be either

— technological, e.g. the lack of technologies for certain post-harvest operations, or

— financial and economic, e.g. for certain required operations or inputs the incurred costs are too high, or

— institutional, e.g. lack of infrastructure and barriers for contractual arrangements.

Most of the constraint hypotheses suggested in the expert meeting relate to the institutional level which led to the assumption that there might be major imperfections in the post-harvest systems of both crops. The constraint hypotheses were guiding the selection of interview partners and study sites, as well as the analysis of survey findings.

The assessment of constraints is strongly related to the performance of the post-harvest system. The performance of systems is usually measured by its productivity, efficiency, the flexibility to survive and the ability to absorb external shocks.

The analysis of the survey data found less constraints in the system than expected. The post-harvest system is generally able to cope with the specific characteristics of root and tuber crops.

The transport system is coping with the bulkiness of both commodities by establishing various lines of transport according to their relative costs. Perishability of the sweet potato is a potential hazard. Although handling is generally rough, the quality of the product offered to final destinations is not affected. Losses are generally seen as negligible.

However, major areas could be identified in which the systems' performance is hampered:

— The state of the rural road infrastructure is increasing transport costs. In the Meru and Nyandarua districts, many roads are poor during rainy periods. In the Kisii and Rachuonyo districts, feeder roads are inaccessible and thus unable to provide access from distant production areas to the main roads, especially during the rains.

— Transparency on the volume traded in the market is low. Packaging is not standardised, which leads to faulty information for actors who are not familiar with marketing. Furthermore, farmers are unable to interpret prices expressed in bags of unknown weight o. volume.

— Potato supply is constrained by the lack of certified seed.

— Traditional eating habits hamper the further commercialisation of sweet potato.

— Research-extension linkages to and from the farmer are weak, especially for sweet potato.

On the technical level, operations like harvesting, grading and storage only require little improvement. On-farm processing of sweet potato for commercial purposes is still in its initial stage and lacks economic attractiveness.

Lack of credit did not appear a major problem for many actors. Although individuals claim the need for it, systematic access barriers to capital could not be found. Entry barriers to markets and operations exist, but are not leading to excessive margins and profits.

A high degree of trust among wholesale market actors was observed, thus indicating the performance of informal contractual arrangements. Systems analysis has been proven as a suitable tool for assessing the performance and identifying constraints for further development. The performance of the post-harvest system of the two crops is quite good. No major inefficiencies at the present stage were observed.

The development perspectives of the two systems are different and depend on various factors. Potato as an already highly commercialised crop has entered the diet of both urban and rural consumers in many areas of Kenya. Commercialisation and supply have steadily developed to a system that shows enough flexibility and diversity to secure a constant supply to market demand.

The capacity of the existing sweet potato post-harvest system is strongly interrelated to the current demand pattern. Urban and rural town demand drives the development. Interventions that aim for an increase in production areas in order to stimulate commercialisation should therefore not be made without prior analysis of market demand development.

Gender participation in post-harvest systems

Potato growing and marketing within the main growing areas is highly commercialised As such it cannot be described as a women's or a man's crop, but a family affair. In the more intensive farming systems such as found in Meru, all family members participate in most of the farm activities and occasionally hire labour when needed. In the Meru area, there was no clear distinction between gender roles in the production of potatoes. In the local markets, there were more women wholesaling and retailing potatoes than men. At the farm level, men and women were active in production, harvesting, sorting and selling.

In the marketing chain, the majority of retailers in the local markets were women, while the men did the wholesale and transportation functions.

Two scenarios were observed in Kisii in relation to gender participation in potato production and sales. In the Mosocho area where sweet potato is highly commercialised both men and women were involved in the production processes and post-harvest systems of grading and selling. It was also observed that women form the majority of traders in both the local and distant markets.

For the distant markets in major towns such as Nakuru, Nairobi and Mombasa, two or three women or a combination of women and men join up and hire a lorry to transport sweet potato to the chosen destination. This may be happening either because one person may not be able to purchase enough produce to fill a whole truck, due to fear of handling large amounts of a highly perishable crop, or due to limited resources among the traders, most of whom are women.

In the Marani division which is further away from the market centres, farmers grow sweet potatoes for home consumption and for local markets. As such it is purely a woman's crop and the men were not interested in the activities related to sweet potatoes.

The entry barriers to both the potato and sweet potato trade by women, may be associated with the initial capital requirements. A beginner requires enough cash to hire a lorry and also to buy the produce. She also has to deal with the cultural expectations as to the role of a woman. Secondly she must struggle to penetrate this male dominated business of transportation.

References

Alumira, J. and C. Obara, *Post-Harvest Consumption Analysis of Sweet Potato in Kenya - Survey*, Report commissioned by GTZ, Nairobi, 1997.

Anand, Nikhil. "Indian Potato Exports: Problems and Prospects", *Journal of the Indian Potato Association*, 2002.

Bibliography

Alumira, J. and C. Obara, *Post-Harvest Consumption Analysis of Sweet Potato in Kenya - Survey*, Report commissioned by GTZ, Nairobi, 1997.

Anand, Nikhil. "Indian Potato Exports: Problems and Prospects", *Journal of the Indian Potato Association*, 2002.

Biaugeaud, H., *Food processing equipment*, Technical Bulletin, Henri Biaugeaud, S.A.: Arcueil, 1994.

Boyle, M., W.T. Frankenberger, Jr., and L.H. Stolzy, "The influence of organic matter on soil aggregation and water infiltration", *Journal of Production Agriculture*, 1989.

Brook, R. C., R. J. Fick, and T. D. Forbush, Potato Storage D*esign and Managemen, East Lansing, MI.: Michigan State University* Agricultural Engineering Information Series, AEIS#627, 1995.

Bruinsma, D.H., Witsenburg, W.W., Wurden, W., *Selection of technology for food processing in developing countries*, Publ of the International Course in Food Science and Nutrition, Centre for Agricultural Publishing and Documentation, Wageningen, The Netherlands, 1983.

Bugg, R. L., and C. Waddington, "Using Cover Crops to Manage Arthropod Pests of Orchards - a Review", *Agriculture Ecosystems & Environment*, 1994.

Burton, W.G., *The potato*, A survey of its history and of factors influencing its yield, nutritive value, quality and storage, edit. H. Veenman & Zonen N.V., Wageningen, The Netherlands, 1966.

Cederbaum, S. B., J. P. Carroll, and R. J. Cooper, "Effects of alternative cotton agriculture on avian and arthropod populations", *Conservation Biology*, 2004.

Celis, C. *et al.*, "Environmental biosafety and transgenic potato in a center of diversity for this crop", *Nature*, 2004.

Christ, B. J., *Potato Diseases in Pennsylvania*, Pennsylvania College of

Agricultural Sciences, University Park, PA. 1998.

Colin, D., "Recent trends in fruit and vegetable processing", *In Food Science and Technology Today,* 1992.

Creamer, *et al,* "Mechanisms of weed suppression in cover crop-based production systems", HortScience, 1996.

Dabney, S. M., J. A. Delgado, and D. W. Reeves, "Using winter cover crops to improve soil quality and water quality", *Communications in Soil Science and Plant Analysis,* 2001.

Diop, A and Wanzie, R., *Improved low-cost storage of potatoes in the North-west province of Cameroon,* Proceeding of the Workshop on the African Experience on Post-harvest Technology Development, Accra (Ghana), 4-8 July 1994.

Ditsch, D. C., and M. M. Alley, "Nonleguminous Cover Crop Management for Residual N Recovery and Subsequent Crop Yields", *Journal of Fertilizer Issues,* 1991.

Dunn, J. W., J. K. Harper, and G. L. Greaser, *AgriculturalAlternatives: Fruit and Vegetable Marketing for Small-scaleand Part-time Growers,* University Park, PA: Penn StateCooperative Extension, 2000.

Food and Agriculture Organization of the United Nations, *Prevention of postharvest food losses: A training manual,* Rome: FAO Training Series, 1985.

Frederick, R.J. Virgin, I. Lindarte, E. (eds.), *Environmental concerns with transgenic plants in center of diversity: Potato as a model, proceedings,* Stockholm (Sweden), Stockholm Environmental Institute (SEI), Biotechnology Advisory Commission (BAC) Inter-American Institute for Cooperation on Agriculture (IICA), 1995.

Freemark, K. E., and D. A. Kirk, "Birds on organic and conventional farms in Ontario: partitioning effects of habitat and practices on species composition and abundance", *Biological Conservation,* 2001.

Galloway, *et al.,* "Nitrogen-Fixation - Anthropogenic Enhancement-Environmental Response", *Global Biogeochemical Cycles,* 1995.

Giller, K. E., and G. Cadisch, "Future benefits from biological nitrogen fixation: An ecological approach to agriculture", *Plant and Soil* (Historical Archive), 1995.

Grafton-Cardwell, *et al.,* "Leguminous cover crops to enhance population development of Euseius tularensis (Acari : Phytoseiidae) in citrus", *Biological Control,* 1999.

Greaser, G. L. and J. K. Harper, *Agricultural Alternatives: Enterprise Budget Analysis,* University Park, PA: Penn StateCooperative Extension, 1994.

Haramoto, E. R., and E. R. Gallandt, "Brassica cover cropping for weed management: A review", *Renewable Agriculture and Food Systems,* 2004.

Hawkes, J.G., *The Potato: Evolution, Biodiversity & Genetic Resources,* Smithsonian Institution Press, Washington, D.C., 1990.

Hunt, G.L.T., *Low cost potato storage for the tropics,* Acta Horticulturae 158, Tropical Horticulture 10, 1985.

Jackson, S.A. Hanneman Jr., R.E., "An assessment of crossability between cultivated potato and its wild relatives", *American Potato Journal (USA),* 1996.

Jordaan, A.J., and Van Schalkwyk, H.D., *Impact of the Potato Industry on the South African Economy,* Department of Agricultural Economics, Free State University, Bloemfontein, 2002.

Kader, A.A., ed., *Post-harvest technology of horticultural crops,* Oakland: University of California, Division of Agriculture and Natural Resources Publication 3311, 2002.

Kitinoja, L. and A.A. Kader, *Small-scale post-harvest handling practices: A manual for horticultural crops,* Fourth edition. Davis: University of California, Postharvest Horticulture Series 8E, 2002.

Lamont, W. J., *et al., Agricultural Alternatives: Irrigation for Fruit and VegetableProduction,* University Park, PA: Penn State CooperativeExtension, 2001.

Lang, James, *Notes of a Potato Watcher,* Texas A&M University, College Station, TX, 2001.

Larry Zuckerman, *Potato, The: How the Humble Spud Rescued the Western World,* Douglas & McIntyre, 1999.

Magee, T.R.A. and Wilkinson, C.P.D., "Influence of process variables on the drying of potato slices", *In Int. Journal of Food Science and Technology,* 1992.

Matz, S.A., "Potato Chips", *In Snack Food Techology,* AVI Publishing Co., Inc. Westport, Conn., 1984.

Mitra, S., ed., *Post-harvest physiology and storage of tropical and subtropical fruits,* Wallingford, UK: CAB International, 1997.

Nahdy, S. M. and Odong, M., " Storage of fresh cassava tuber in plant based storage media", In: Proceeding of the Workshop on *Post-harvest Technology Experience in Africa,* Accra, 4-8 July 1994.

Pimentel, D. *et al.,* "Environmental and economic costs of soil erosion and conservation benefits", *Science,* 1995.

Potatoes South Africa, *The South African potato industry. A review,* Potatoes South Africa, Pretoria, 1999.

Pushkarnath, *Potato in sub-tropics*, Editor: Orient Longman, 1976.

Rastovski, A., Van Es, A. *et al.*, *Storage of potatoes: post-harvest behaviour, store design, storage practice, handling*, Centre for Agricultural Publishing and Documentation, Wageningen, 1981.

Rowe, Randall, editor, *Potato Health Management*, St. Paul,MN: APS Press, 1993.

Salaman, Redcliffe N., *The History and Social Influence of the Potato*, Cambridge University Press, 1989.

Shaw, R. and Booth, R. (undated), *Simple Processing of Dehydrated Potatoes and Potato Starch*, International Potato Centre (CIP), Lima, Peru, 1981.

Shewfelt, R.L. and S.E. Prussia, eds., *Post-harvest handling: A systems approach*, San Diego, CA: Academic Press, 1993.

Sieczka, J. B. and R. E. Thornton, *Commercial PotatoProduction in North America*, Orono, ME: The PotatoAssociation of America Handbook, 1992.

Soil Association *Soil Management on Organic Farms*, Available from: Soil Association, 2003.

Stevenson, W.R., *et al.*, *Compendium of Potato Diseases*, 2nd ed, Amer. Phytopathological Society, St. Paul, MN., 2001.

Thompson, A.K., "Post-harvest technology of fruits and vegetables", Oxford: Blackwell Science, 1996.

Thompson, J.F., A.A. Kader, and K. Sylva, *Compatibility chart for fruits and vegetables in short-term transport or storage*, Oakland: Univ. Calif. Div. Agric. Nat. Res. Publ. 21560 (poster), 1996.

US Department of Agriculture, *Soil Biodiversity*, Soil Quality Information Sheet, Soil Quality Resource Concerns, January 1998.

Weisz, R., Z. Smilowitz, M. Saunders, and B. Christ, *Integrated Pest Management for Potatoes*, University Park,PA: Penn State Cooperative Extension, 1995.

Werge RW, "Potato Processing in the Central Highlands of Peru", *Ecology of Food and Nutrition*, food processing, potato; South America; Peru, 1979.

Wills, R., B. McGlasson, D. Graham, and D. Joyce, *Postharvest: An introduction to the physiology and handling of fruit, vegetables and ornamentals*, Wallingford: CAB International, 1998.

Index